the definitive guide to
BETTING ON THE ALL-WEATHER

RACING POST
expertseries

6

the definitive guide to

BETTING ON THE ALL-WEATHER

David Bellingham

Raceform

To Clare

Many thanks to Julian Brown, Richard Lowther, Tracey Scarlett and Steffan Edwards.

Some of the statistics contained within this book were compiled with the assistance of Raceform Interactive.

In-running betting details reproduced by kind permission of Betfair.

Published in 2008 by Racing Post,
Axis House, Compton, Newbury, Berkshire, RG20 6NL

A catalogue record for this book is available from the British Library.

ISBN 978-1-905153-93-0

Designed by Tracey Scarlett

Printed by CPI William Clowes Beccles NR34 7TL

CONTENTS

INTRODUCTION

TIME DOES not stand still – take my word for it. In the year 2000 my first book, called *Gold From The Sand*, was published and contained within it was every possible piece of meaningful information I could think of about the three All-Weather courses in existence at the time, namely the Lingfield Equitrack plus the Fibresand surfaces at Southwell and Wolverhampton.

However, within a year of the book coming out Lingfield decided to dig up its Equitrack surface and replace it with Polytrack, and then in the autumn of 2004 Wolverhampton changed from Fibresand to Polytrack. An updated version of *Gold From The Sand* was therefore necessary, which duly saw the light of day in the autumn of 2005.

Within six months of its publication, everything changed once again as Kempton opened up its new right-handed Polytrack surface. Then in August 2007 Dundalk became Ireland's first All-Weather track (with all due respect to Laytown) and then, most recently and after many frustrating delays, Great Leighs finally got under way in the April of 2008 to become the sixth All-Weather track in the British Isles.

An updated reference covering the six sand tracks within these islands and how to tackle the races run on them again became necessary, and here it is. Hopefully this volume will remain pertinent for a decent amount of time before any more sand tracks come online, but you never can tell.

The book starts off with a chapter dedicated to each of the six sand tracks we currently have. Each contains a brief description of the venue, but I have deliberately kept these short as the most meaningful section of these chapters will be the statistics that cover jockeys, trainers, favourites, sires, the draw and the record of front-runners. Obviously with Dundalk and Great Leighs having only recently come online in relative terms, the statistics for them will be of more limited significance.

Pace and track biases and how to spot them and capitalise on them are covered, as are the more routine subjects of form analysis and how the shape of a race can have a major impact on the result. There is also a chapter dedicated to ratings and within that is a blow-by-blow description of how I have developed and maintain my own speed ratings.

Two brand new subjects are also covered. In the chapter on track biases is a section demonstrating how betting-in-running can be a powerful tool for those willing to keep their eyes open, whilst sectional times has a chapter all to itself despite its all-too-brief appearance in the public domain.

An attempt has been made to try and assemble all the pieces previously described in the chapter appropriately called "Putting It All Together", whilst the book ends with a diary covering each of the nine All-Weather meetings I attended between March and July 2008 in which I used many of the tools covered in these pages, and how my betting bank fared as a result.

I sincerely hope that you enjoy reading this book, as quite apart from it being my job and an opportunity to make some extra money through betting, following horse racing should also be fun. I certainly had a lot of fun putting this volume together.

LINGFIELD

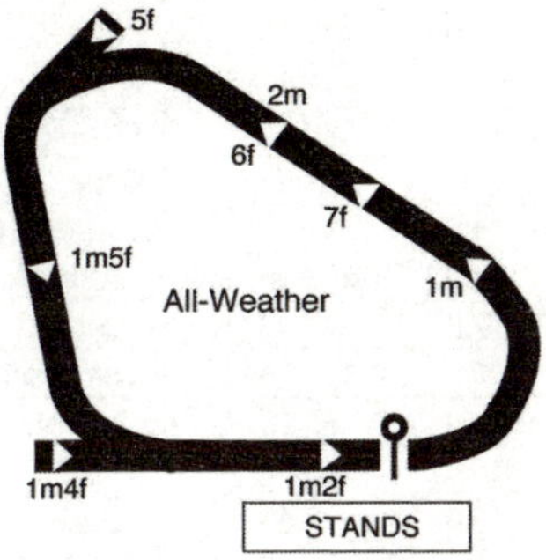

LINGFIELD was the first track to stage an All-Weather meeting in Britain (October 30th 1989) and in those days they raced on Equitrack. However, after one or two problems with the surface it was replaced by Polytrack in November 2001, and it wasn't long before the quality of horse appearing at the venue rose significantly.

Another major step forward for All-Weather racing took place here in March 1998 with the introduction of the Winter Derby, then just a conditions event but now a Group 3 contest. The inaugural running was won by Running Stag and ever since then the race has been regarded as the 'Blue Riband' of sand racing in this country. There are several Listed races run here during the course of the year too.

Lingfield differs from the other five All-Weather tracks in the British Isles in that it is basically a triangular-shaped circuit of nearly ten furlongs with a sweeping turn into a home straight of little more than a furlong, the shortest of all the sand tracks. That configuration creates its own problems, with racing room at such a premium, and it's probably fair to say that there is more trouble in running here than just about anywhere else in the country, sand or otherwise.

As a result on many occasions the best horse doesn't win and it's well worth making a note of those that are denied a run but for which they would have gone close to winning. Also keep an eye on how horses fare when they stick tight against the inside rail once into the home straight, as in recent years there have been long periods when horses taking this route have been at a major disadvantage. Either they have made a promising-looking effort along the rail only to then flatten out, or they have stayed there after leading into the straight and been caught.

There was also a long period when races run over trips of a mile and a half or further were run at an early dawdle, probably because the jockeys got it into their heads that it wasn't possible to make all over the longer distances here. Certainly the statistics for front-runners are better for trips up to a mile, but it is possible to race handily and win over further if the pace is judged right.

Lingfield Statistics July 2005 to July 2008

JOCKEYS

	TOTAL W-R	PER CENT	£1LEVEL STAKE
Seb Sanders	68-388	18.0	-36.91
George Baker	63-377	17.0	+27.59
Jamie Spencer	60-316	19.0	-32.18
N Callan	59-362	16.0	+11.77
Eddie Ahern	58-492	12.0	-60.32
Dane O'Neill	56-504	11.0	-81.73
John Egan	51-382	13.0	-47.37
Chris Catlin	51-605	8.0	-209.06
Richard Hughes	50-291	17.0	-27.98
Steve Drowne	46-466	10.0	-123.76
James Doyle	41-387	11.0	-65.18
Ryan Moore	39-237	16.0	-64.36
Joe Fanning	32-216	15.0	-73.48
Ian Mongan	32-285	11.0	+25.99
Martin Dwyer	28-295	9.0	-87.92
Hayley Turner	28-363	8.0	-145.15
Jim Crowley	27-300	9.0	-93.28
L P Keniry	27-369	7.0	-70.60
Jimmy Quinn	24-335	7.0	-96.17
Richard Kingscote	24-126	19.0	+15.46
Adam Kirby	24-289	8.0	-98.88

SIRES

	TOTAL W-R	PER CENT	£1 LEVEL STAKE
Green Desert	40-179	22.3	+50.85
Dansili	34-228	14.9	-13.38
Singspiel	31-154	20.1	+39.25
Grand Lodge	29-177	16.4	+45.25
Pivotal	27-170	15.9	-21.40
Cape Cross	24-180	13.3	+35.81
Cadeaux Genereux	24-188	12.8	-14.75
Atraf	24-117	20.5	+55.49
Mark Of Esteem	22-210	10.5	-69.50
Compton Place	21-223	9.4	-47.01
Danehill Dancer	21-130	16.2	-14.58

FAVOURITES

	W-R	PER CENT	£1 LEVEL STAKE
2yo Stks	85-252	33.7	-28.74
3yo Stks	62-142	43.7	+22.27
3yo+ Stks	125-411	30.4	-72.01
Totals	**272-805**	**33.8**	**-78.48**
2yo H'caps	14-51	27.5	-11.32
3yo H'caps	70-195	35.9	+32.11
3yo+ H'caps	194-796	24.4	-68.36
Totals	**278-1042**	**26.7**	**-47.57**
Apprentice	14-59	23.7	-9.55
Amateurs	6-12	50.0	+12.88
Totals	**20-71**	**28.2**	**+3.33**
All Favs	570-1918	29.7	-122.72

Lingfield All-Weather Track

TRAINERS

	TOTAL W-R	2YO STKS	3YO STKS	OTHER STKS	2YO H'CPS	3YO H'CPS	OTHER H'CPS	APP	AMAT	PER CENT	£1 LEVEL STAKE
R Hannon	59-386	19-92	4-37	8-60	2-24	11-54	14-115	1-4	0-0	15.3	-65.39
G L Moore	50-362	3-17	1-10	19-84	0-3	0-8	23-208	2-22	2-10	13.8	+12.90
J R Best	38-363	4-45	4-28	4-44	2-16	5-46	18-168	1-14	0-2	10.5	-132.83
N P Littmoden	33-272	1-17	0-11	5-60	1-6	8-41	16-127	2-8	0-2	12.1	-75.20
M Johnston	26-200	5-43	4-39	9-33	0-12	7-38	1-30	0-5	0-0	13.0	-98.35
D Shaw	26-198	1-9	0-1	2-27	0-0	4-20	17-131	2-7	0-3	13.1	+23.79
T G Mills	26-104	6-22	5-11	1-13	2-4	6-21	6-32	0-1	0-0	25.0	+29.77
M A Jarvis	25-105	6-21	4-14	7-26	1-3	2-8	5-33	0-0	0-0	23.8	+34.39
Sir M Prescott	24-112	0-37	5-10	6-17	1-4	3-19	9-23	0-2	0-0	21.4	-4.54
J S Moore	23-158	3-47	1-5	5-32	1-5	4-22	8-40	0-5	1-2	14.6	+13.25
P J Makin	22-119	5-20	0-7	1-20	3-4	2-14	11-53	0-1	0-0	18.5	-16.34
M P Tregoning	22-101	4-32	6-16	9-30	0-3	2-10	1-10	0-0	0-0	21.8	-5.38
A M Balding	22-159	1-21	5-19	5-35	0-6	5-28	6-44	0-6	0-0	13.8	-37.77
C E Brittain	21-187	3-28	3-14	6-40	0-5	2-28	7-67	0-5	0-0	11.2	-42.81
E A L Dunlop	21-141	10-44	2-23	1-25	1-5	6-26	1-18	0-0	0-0	14.9	+21.48
J Noseda	21-94	7-28	3-16	8-30	0-1	1-10	2-9	0-0	0-0	22.3	-9.91
J A Osborne	20-135	4-30	4-35	4-23	1-5	5-29	1-12	1-1	0-0	14.8	+14.28
D Elsworth	19-152	5-30	2-23	2-34	0-3	2-13	7-46	1-1	0-2	12.5	+21.50
P F I Cole	19-131	4-27	1-9	5-29	1-4	3-17	5-40	0-5	0-0	14.5	-9.30
K A Ryan	19-112	4-13	0-7	5-22	0-5	1-9	8-54	0-0	1-2	17.0	-10.90
M R Channon	19-161	8-28	1-26	1-22	1-10	3-36	5-36	0-3	0-0	11.8	-15.80

Lingfield Draw and Front-Runner Statistics July 2005 to July 2008

5 FURLONGS

Draw Pos	Winners	%
Low	34	25
Middle	58	43
High	44	32

DRAW NO	RUNNERS	WINNERS	%
1	131	14	11
2	128	11	9
3	130	20	15
4	133	18	14
5	127	19	15
6	122	14	11
7	109	12	11
8	104	10	10
9	85	11	13
10	73	7	10
Highest	136	20	15

Front Runners Impact Value = 1.9
37 winners from 136 races (27%)

6 FURLONGS

Draw Pos	Winners	%
Low	101	32
Middle	117	38
High	93	30

DRAW NO	RUNNERS	WINNERS	%
1	290	25	9
2	291	42	14
3	291	42	14
4	299	30	10
5	293	25	9
6	288	32	11
7	283	25	9
8	257	28	11
9	230	21	9
10	195	19	10
11	167	13	8
12	135	9	7
Highest	311	41	13

Front Runners Impact Value = 1.4
47 winners from 311 races (15%)

7 FURLONGS

Draw Pos	Winners	%
Low	87	24
Middle	145	40
High	127	35

DRAW NO	RUNNERS	WINNERS	%
1	342	28	8
2	343	25	7
3	340	31	9
4	341	33	10
5	338	35	10
6	328	36	11
7	309	35	11
8	289	39	13
9	274	24	9
10	254	22	9
11	221	17	8
12	190	15	8
13	150	9	6
14	116	10	9
Highest	359	64	18

Front Runners Impact Value = 1.5
55 winners from 359 races (15%)

1 MILE

Draw Pos	Winners	%
Low	90	28
Middle	136	42
High	101	31

DRAW NO	RUNNERS	WINNERS	%
1	311	30	10
2	318	38	12
3	303	36	12
4	312	33	11
5	303	31	10
6	297	46	15
7	281	30	11
8	266	19	7
9	243	14	6
10	225	21	9
11	196	10	5
12	168	19	11
Highest	327	43	13

Front Runners Impact Value = 1.4
50 winners from 327 races (15%)

1 MILE 2 FURLONGS

Draw Pos	Winners	%
Low	79	26
Middle	124	41
High	102	33

DRAW NO	RUNNERS	WINNERS	%
1	291	25	9
2	286	28	10
3	288	23	8
4	286	39	14
5	293	31	11
6	284	28	10
7	275	39	14
8	248	21	8
9	211	18	9
10	194	16	8
11	171	10	6
12	145	9	6
13	124	7	6
14	99	11	11
Highest	305	49	16

Front Runners Impact Value = 0.9
27 winners from 305 races (9%)

1 MILE 4 FURLONGS

Draw Pos	Winners	%
Low	41	25
Middle	62	38
High	62	38

DRAW NO	RUNNERS	WINNERS	%
1	160	12	8
2	147	16	11
3	156	13	8
4	151	16	11
5	156	19	12
6	151	24	16
7	130	15	12
8	130	14	11
9	109	9	8
10	90	5	6
11	74	8	11
12	62	4	6
13	58	5	9
14	49	4	8
15	38	1	3
16	35	0	0
Highest	165	27	16

Front Runners Impact Value = 0.7
13 winners from 165 races (8%)

1 MILE 5 FURLONGS

Draw Pos	Winners	%
Low	9	21
Middle	14	33
High	20	47

DRAW NO	RUNNERS	WINNERS	%
1	40	6	15
2	40	1	3
3	43	5	12
4	41	0	0
5	40	7	18
6	38	3	8
7	38	8	21
8	33	3	9
9	28	1	4
10	26	3	12
11	18	1	6
12	15	2	13
13	12	2	17
14	9	1	11
Highest	43	9	21

Front Runners Impact Value = 0.0
0 winners from 43 races (0%)

2 MILES

Draw Pos	Winners	%
Low	7	18
Middle	20	51
High	12	31

DRAW NO	RUNNERS	WINNERS	%
1	38	0	0
2	37	6	16
3	36	3	8
4	38	4	11
5	36	3	8
6	35	4	11
7	35	5	14
8	32	3	9
9	30	4	13
10	27	2	7
11	21	1	5
12	17	0	0
13	16	4	25
14	17	0	0
Highest	39	7	18

Front Runners Impact Value = 0.7
3 winners from 39 races (8%)

The Lingfield draw statistics suggest that there are few biases at the track these days, certainly less than on the old Equitrack. You can win from just about anywhere over five furlongs which is interesting when you consider the two sharp left-handed bends the runners have to encounter. Early pace is a different matter, however, with early leaders winning almost twice as often as they are entitled to (impact value 1.9).

Over trips ranging from six furlongs to a mile, a middle draw seems to be an advantage and prominent racers are marginally favoured, but the draw advantage swings towards those drawn low over a mile and a quarter and a mile and a half. Note also that it becomes difficult, though not impossible, to lead early and win races over the longer trips.

SOUTHWELL

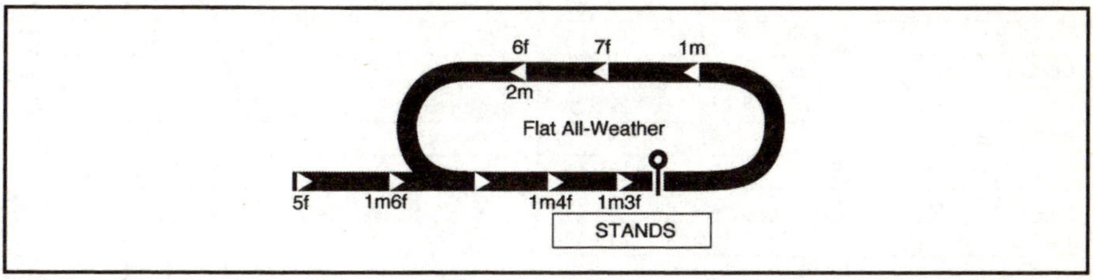

TOGETHER with Lingfield, Southwell is one of the two original British All-Weather tracks (first meeting November 8th 1989) but unlike the Surrey venue, Southwell has stuck with its original surface of Fibresand throughout.

However, things very nearly went permanently pear-shaped for the Nottinghamshire track in June 2007 when persistent rain caused the nearby River Greet to burst its banks, leaving the track and the racecourse buildings under two feet of water. When the floods finally subsided, there was the amazing spectacle of dead fish lying on the ground floor of the grandstand and there were fears that the course may never reopen at all given the clean-up bill involved. Even when the track was given the green light to reopen there were calls from some quarters for the course to switch to Polytrack.

From a personal point of view, I'm glad they stuck with Fibresand as so many horses clearly prefer it and with five other tracks now staging racing on Polytrack in the UK and Ireland, there is hardly a lack of opportunities for horses on that surface. I already liked the venue even before the disaster struck, but following its revamp I love it even more and the facilities for what is a minor track are truly excellent in my view. Also I just seem to do better punting-wise there than anywhere else!

The circuit is around ten furlongs in circumference but with a straight five-furlong track, which is unique for an All-Weather venue in these islands. The run-in from the home bend on the round course is almost three furlongs, and it is well worth keeping an eye out for any track biases. If you do find yourself racing on the slowest part of the track here, the surface will really find you out and for quite a long period it was an absolute no-no to stick close to the inside rail down the home straight. It doesn't seem to be so much the case at present, but that doesn't mean it won't happen again and we should be ready to capitalise on it if it does.

The Fibresand surface is much more testing than Polytrack and conventional wisdom suggests that it should suit horses that like a bit of cut in the ground on turf, but that is a dangerous assumption as many horses have hated their first encounter with Fibresand whatever their preferences on grass. It's a theme that crops up on a few more occasions in this book, but the advice is to concentrate on horses that have already shown a liking for Fibresand in the past.

Southwell Statistics July 2005 to July 2008

JOCKEYS

	TOTAL W-R	PER CENT	£1 LEVEL STAKE
Chris Catlin	47-398	12.0	-23.80
N Callan	42-330	13.0	-113.85
Micky Fenton	31-299	10.0	+44.66
Hayley Turner	31-267	12.0	-24.27
Phillip Makin	31-206	15.0	-58.58
Seb Sanders	28-112	25.0	+0.63
Robert Winston	28-236	12.0	-44.59
T P Queally	26-136	19.0	-10.64
Pat Cosgrave	25-192	13.0	-26.22
R Fitzpatrick	24-175	14.0	-5.04
Paul Hanagan	24-229	10.0	-113.83
Daniel Tudhope	24-155	15.0	-12.92
Dale Gibson	23-278	8.0	+18.66
Joe Fanning	23-137	17.0	+20.72
Greg Fairley	23-113	20.0	+12.16
Tony Culhane	22-196	11.0	-77.76
S Donohoe	22-152	14.0	-11.30
Tom Eaves	18-179	10.0	-17.63
J-P Guillambert	18-160	11.0	-56.01
Robert Havlin	17-121	14.0	+98.02
Graham Gibbons	17-163	10.0	-51.58

SIRES

	TOTAL W-R	PER CENT	£1 LEVEL STAKE
Diktat	23-148	15.5	-14.54
Orpen	20-157	12.7	-0.05
Tagula	18-107	16.8	+0.48
Mujahid	17-105	16.2	+31.99
Bertolini	17-129	13.2	+14.12
Medicean	15-62	24.2	+70.47
Dr Fong	14-75	18.7	+31.80
Vettori	14-135	10.4	-41.17
Lujain	14-119	11.8	-30.62
Komaite	13-127	10.2	-44.75
Fasliyev	13-90	14.4	-31.02
Dancing Spree	13-70	18.6	+48.67
Piccolo	13-151	8.6	-20.67

FAVOURITES

	W-R	PER CENT	£1 LEVEL STAKE
2yo Stks	30-79	38.0	-6.94
3yo Stks	30-81	37.0	-1.48
3yo+ Stks	158-480	32.9	-8.94
Totals	**218-640**	**34.1**	**-17.36**
2yo H'caps	8-31	25.8	-3.84
3yo H'caps	32-99	32.3	-12.50
3yo+ H'caps	168-571	29.4	-3.07
Totals	**208-701**	**29.7**	**-19.41**
Apprentice	8-41	19.5	-15.57
Amateurs	3-19	15.8	-7.75
Totals	**1-60**	**18.3**	**-23.32**
All Favs	**437-1401**	**31.2**	**-60.09**

Southwell All-Weather Track

TRAINERS

	TOTAL W-R	2YO STKS	3YO STKS	OTHER STKS	2YO H'CPS	3YO H'CPS	OTHER H'CPS	APP	AMAT	PER CENT	£1 LEVEL STAKE
P A Blockley	35-221	1-13	1-17	13-92	1-3	3-9	13-78	2-5	1-4	15.8	-37.30
K A Ryan	30-229	4-32	4-21	9-41	1-11	3-21	8-98	1-3	0-2	13.1	-88.07
R A Fahey	28-185	1-11	0-4	7-50	1-4	1-9	17-100	1-7	0-0	15.1	-31.34
M Johnston	26-86	3-14	3-12	5-17	1-5	8-20	6-18	0-0	0-0	30.2	+43.80
A J McCabe	23-193	2-6	2-13	4-43	1-4	1-23	13-98	0-5	0-1	11.9	-47.72
J Balding	21-150	0-8	0-9	11-71	0-0	0-3	10-58	0-1	0-0	14.0	+38.24
D Carroll	21-145	0-3	1-10	8-61	0-1	0-5	12-61	0-3	0-1	14.5	-6.79
Sir M Prescott	20-71	1-9	1-10	2-12	3-6	4-9	8-23	1-1	0-1	28.2	-4.25
T D Barron	20-129	1-3	2-10	3-32	0-4	2-18	11-60	0-1	1-1	15.5	-38.73
P Howling	19-196	0-1	0-7	13-108	0-2	0-8	6-65	0-3	0-2	9.7	-38.00
R Hollinshead	19-152	0-5	2-9	10-54	2-4	0-10	4-64	1-5	0-1	12.5	+40.26
P W Hiatt	19-130	0-1	0-2	7-52	0-0	0-2	10-63	1-7	1-3	14.6	-7.06
H Morrison	19-79	3-6	2-3	6-11	0-0	2-13	6-45	0-1	0-0	24.1	+37.50
K R Burke	18-153	0-7	2-15	8-60	0-3	0-13	7-50	0-3	1-2	11.8	-63.74
D Shaw	18-240	0-6	0-11	4-58	1-12	2-22	10-129	0-1	1-1	7.5	-66.63
J R Boyle	18-96	0-4	6-10	5-37	0-0	1-4	6-36	0-3	0-2	18.8	-10.34
Peter Grayson	17-105	0-9	1-9	3-17	2-7	3-17	7-42	1-3	0-1	16.2	+19.20
J G Given	16-89	2-10	2-11	5-36	0-1	3-9	4-21	0-1	0-0	18.0	+26.53
J A Osborne	16-64	4-5	3-11	5-27	0-1	2-6	2-13	0-1	0-0	25.0	-6.21
M J Attwater	16-154	0-6	1-5	12-83	0-1	1-8	2-46	0-2	0-3	10.4	-43.42
S R Bowring	14-169	1-9	0-5	7-84	0-0	1-6	5-59	0-5	0-1	8.3	-83.92

Southwell Draw and Front-Runner Statistics July 2005 to July 2008

5 FURLONGS

Draw Pos	Winners	%
Low	65	38
Middle	72	42
High	34	20

DRAW NO	RUNNERS	WINNERS	%
1	159	29	18
2	164	24	15
3	161	17	11
4	162	27	17
5	161	21	13
6	157	16	10
7	140	10	7
8	125	6	5
9	110	8	7
10	93	4	4
11	70	4	6
12	61	3	5
13	46	2	4
14	38	0	0
Highest	171	14	8

Front Runners Impact Value = 2.4
43 winners from 171 races (25%)

6 FURLONGS

Draw	Pos	%
Low	76	31
Middle	103	42
High	66	27

DRAW NO	RUNNERS	WINNERS	%
1	232	22	9
2	231	25	11
3	226	20	9
4	230	40	17
5	240	32	13
6	225	19	8
7	216	21	10
8	197	15	8
9	176	11	6
10	152	19	13
11	125	12	10
12	100	2	2
13	80	5	6
14	66	2	3
Highest	245	30	12

Front Runners Impact Value = 2.1
54 winners from 245 races (22%)

7 FURLONGS

Draw Pos	Winners	%
Low	54	25
Middle	97	44
High	67	31

DRAW NO	RUNNERS	WINNERS	%
1	209	16	8
2	208	21	10
3	211	22	10
4	208	26	13
5	206	21	10
6	199	25	13
7	185	14	8
8	172	17	10
9	156	20	13
10	144	6	4
11	120	11	9
12	101	7	7
13	80	8	10
14	63	4	6
Highest	218	32	15

Front Runners Impact Value = 1.5
32 winners from 218 races (15%)

1 MILE

Draw Pos	Winners	%
Low	50	18
Middle	130	46
High	105	37

DRAW NO	RUNNERS	WINNERS	%
1	263	9	3
2	266	24	9
3	276	29	11
4	271	28	10
5	262	37	14
6	266	31	12
7	242	33	14
8	225	21	9
9	192	17	9
10	169	9	5
11	143	16	11
12	136	11	8
13	104	10	10
14	95	10	11
Highest	285	51	18

Front Runners Impact Value = 1.1
35 winners from 285 races (12%)

1 MILE 3 FURLONGS

Draw Pos	Winners	%
Low	27	27
Middle	40	40
High	34	34

DRAW NO	RUNNERS	WINNERS	%
1	96	10	10
2	97	14	14
3	96	5	5
4	96	15	16
5	90	7	8
6	91	9	10
7	84	6	7
8	70	5	7
9	68	9	13
10	59	8	14
11	47	4	9
12	41	4	10
13	33	1	3
14	31	4	13
Highest	101	17	17

Front Runners Impact Value = 1.1
12 winners from 101 races (12%)

1 MILE 4 FURLONGS

Draw Pos	Winners	%
Low	21	18
Middle	48	41
High	47	41

DRAW NO	RUNNERS	WINNERS	%
1	111	10	9
2	112	10	9
3	112	12	11
4	109	8	7
5	112	16	14
6	101	14	14
7	90	7	8
8	78	9	12
9	65	11	17
10	57	7	12
11	37	3	8
12	36	4	11
13	30	2	7
14	25	3	12
Highest	116	27	23

Front Runners Impact Value = 1.1
15 winners from 116 races (13%)

1 MILE 6 FURLONGS

Draw Pos	Winners	%
Low	12	22
Middle	22	41
High	20	37

DRAW NO	RUNNERS	WINNERS	%
1	54	4	7
2	54	6	11
3	48	8	17
4	52	3	6
5	51	8	16
6	47	4	9
7	44	5	11
8	39	5	13
9	31	3	10
10	25	5	20
11	22	1	5
12	16	0	0
13	14	2	14
14	12	0	0
Highest	54	10	19

Front Runners Impact Value = 0.5
3 winners from 54 races (6%)

2 MILES

Draw Pos	Winners	%
Low	7	23
Middle	10	33
High	13	43

DRAW NO	RUNNERS	WINNERS	%
1	29	5	17
2	27	1	4
3	30	3	10
4	27	2	7
5	27	3	11
6	30	4	13
7	27	4	15
8	20	5	25
9	17	1	6
10	14	1	7
11	7	0	0
12	5	0	0
13	4	0	0
14	2	1	50
Highest	30	8	27

Front Runners Impact Value = 0.8
3 winners from 30 races (10%)

Southwell's draw and front-runner statistics expose a bias that is rather easier to spot than at Lingfield.

The biggest trend is over the straight five-furlong track and a glance at the statistics will show just what an advantage a low-to-middle draw is, as for many years the centre of the track has ridden faster than elsewhere and that is usually the route the horses drawn in those stalls will take. Early pace is equally if not even more important and if a horse has a combination of that plus a low draw (and a little ability) then he or she becomes a very tempting bet.

The ability to race handily remains important over six furlongs and although the draw is less important than over five, a very high draw is a problem. The bias towards prominently ridden horses decreases in terms of impact value as race distances increase, but it's noticeable that in relative terms they still fare better over middle distances than at Lingfield (the impact value is marginally greater than 1.0).

A true Fibresand specialist; Kylkenny was successful 17 times at Southwell between January 2001 and November 2008

WOLVERHAMPTON

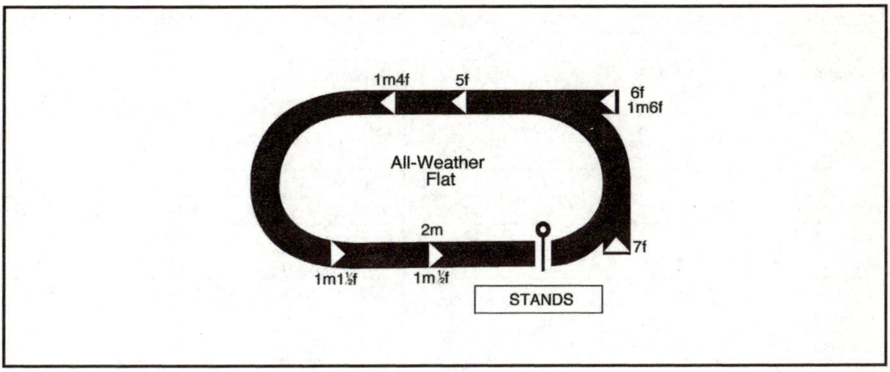

AFTER a long history of staging Flat and National Hunt racing on turf, Wolverhampton opened its doors as an All-Weather track in December 1993 and immediately became the first course in Britain to stage a race under floodlights. The track started off racing on Fibresand and for several years boasted one of the biggest track biases in the country, as any horse that stuck to the inside of the circuit was at a huge disadvantage and you basically knew you had done your money if you saw your selection heading there.

Wolverhampton again made history in December 1995 when it introduced the £50,000 Wulfrun Stakes, the first Listed race to be run on sand in this country, but the loss of the lucrative Saturday evening fixtures meant that savings had to be made and the race was scrapped after 1998. The venue made the switch to Polytrack in October 2004 and as with Lingfield the general quality of the horses racing there improved.

The Wolverhampton circuit is the tightest of the six All-Weather tracks in the British Isles at just under a mile round, but there doesn't appear to be a major track bias at most meetings and horses that stick close to the inside rail don't seem to be that inconvenienced. Even so, it's usually not a good idea to come too wide into the home straight (which is about a furlong and a half in length) because of the ground loss involved.

Wolverhampton Statistics July 2005 to July 2008

JOCKEYS

	TOTAL W-R	PER CENT	£1 LEVEL STAKE
N Callan	85-439	19.0	+55.52
Jamie Spencer	76-286	27.0	+5.25
Eddie Ahern	62-405	15.0	-58.37
Chris Catlin	58-705	8.0	-230.51
Jimmy Quinn	52-508	10.0	-3.25
Seb Sanders	45-299	15.0	-97.81
George Baker	41-279	15.0	+92.14
T P Queally	39-325	12.0	-53.15
Adam Kirby	39-355	11.0	-72.84
Tony Culhane	35-319	11.0	-64.11
Joe Fanning	35-190	18.0	+32.70
Graham Gibbons	35-310	11.0	+49.51
James Doyle	34-350	10.0	-119.41
Francis Norton	32-324	10.0	-38.59
Pat Cosgrave	32-264	12.0	-137.71
Phillip Makin	32-210	15.0	-9.97
Tom Eaves	31-328	9.0	-92.97
Daniel Tudhope	31-226	14.0	-14.64
Liam Jones	31-387	8.0	-129.96
Dean McKeown	30-397	8.0	-137.25
Dane O'Neill	30-284	11.0	-72.27

SIRES

	TOTAL W-R	PER CENT	£1 LEVEL STAKE
Bertolini	45-316	14.2	+120.74
Cadeaux Genereux	37-264	14.0	+4.90
Danetime	36-210	17.1	+44.56
Piccolo	36-266	13.5	-23.65
Desert Sun	29-162	17.9	+108.95
Dansili	26-194	13.4	+2.29
Compton Place	25-211	11.8	-24.00
Pivotal	24-172	14.0	-65.07
Dr Fong	24-196	12.2	-47.75
Green Desert	24-189	12.7	-42.25
Desert Prince	24-182	13.2	-50.53
Tagula	24-217	11.1	-3.07
Spectrum	24-205	11.7	+6.71

FAVOURITES

	PER W-R	£1 CENT	LEVEL STAKE
2yo Stks	104-263	39.5	-11.43
3yo Stks	53-142	37.3	-4.25
3yo+ Stks	194-605	32.1	-70.01
Totals	**351-1010**	**34.8**	**-85.69**
2yo H'caps	27-93	29.0	-4.90
3yo H'caps	53-196	27.0	-24.44
3yo+ H'caps	263-1076	24.4	-160.56
Totals	**343-1365**	**25.1**	**-189.90**
Apprentice	23-77	29.9	+6.92
Amateurs	13-59	22.0	-7.29
Totals	**36-136**	**26.5**	**-0.37**
All Favs	730-2511	29.1	-275.96

Wolverhampton All-Weather Track

TRAINERS

	TOTAL W-R	2YO STKS	3YO STKS	OTHER STKS	2YO H'CPS	3YO H'CPS	OTHER H'CPS	APP	AMAT	PER CENT	£1 LEVEL STAKE
P D Evans	54-394	6-58	6-27	16-94	1-20	6-25	15-143	4-16	0-11	13.7	+6.29
K A Ryan	54-393	8-65	4-32	15-92	3-22	3-32	19-134	0-4	2-12	13.7	-100.74
W Brisbourne	51-548	1-16	1-15	18-173	1-8	1-19	26-277	1-25	2-15	9.3	-221.97
M Johnston	45-201	10-51	6-25	10-32	7-21	8-38	4-29	0-4	0-1	22.4	+26.87
D Shaw	43-456	2-31	2-27	11-107	0-11	3-34	23-229	0-6	2-11	9.4	-91.30
K R Burke	39-280	3-45	4-26	18-80	0-9	1-12	11-97	1-5	1-6	13.9	-74.16
R Hollinshead	38-388	3-34	3-21	15-134	2-14	1-28	14-145	0-5	0-7	9.8	+0.27
R A Harris	33-374	2-40	2-28	7-104	4-13	4-29	11-138	2-15	1-7	8.8	-88.74
W J Haggas	27-119	13-34	1-14	4-24	0-5	1-12	8-30	0-0	0-0	22.7	+2.88
I Semple	27-168	2-16	1-6	4-41	2-5	2-10	16-89	0-1	0-0	16.1	+12.19
J R Boyle	27-135	4-14	4-12	6-33	0-2	4-13	8-49	1-8	0-4	20.0	+18.20
Sir M Prescott	26-122	10-52	3-11	5-19	1-8	2-14	5-18	0-0	0-0	21.3	-20.96
J A Osborne	24-165	8-55	4-27	9-35	2-11	0-20	1-16	0-0	0-1	14.5	-51.28
S Kirk	24-141	8-50	1-12	2-21	3-14	8-27	2-17	0-0	0-0	17.0	+13.56
J Pearce	23-182	1-4	0-5	4-38	0-3	2-13	11-93	1-6	4-20	12.6	-21.37
B P J Baugh	23-267	1-17	0-16	8-87	0-3	0-13	13-111	0-13	1-7	8.6	-37.13
N P Littmoden	23-192	2-28	2-16	5-41	1-8	2-17	9-71	1-6	1-5	12.0	-48.06
P Howling	21-286	3-10	0-13	3-114	0-4	0-5	14-123	1-12	0-5	7.3	-112.05
T D Barron	21-105	1-8	1-10	6-22	1-6	2-8	10-45	0-5	0-1	20.0	+22.00
W J Musson	20-117	0-7	1-3	5-27	0-4	4-10	10-60	0-5	0-1	17.1	+40.13
J W Hills	19-125	4-22	2-11	4-22	3-9	2-19	4-38	0-4	0-0	15.2	-16.75

Wolverhampton Draw and Front-Runner Statistics July 2005 to July 2008

5 FURLONGS 20 YARDS

Draw Pos	Winners	%
Low	89	34
Middle	94	36
High	75	29

DRAW NO	RUNNERS	WINNERS	%
1	251	36	14
2	246	33	13
3	243	28	12
4	242	25	10
5	236	25	11
6	226	27	12
7	213	25	12
8	198	14	7
9	187	21	11
10	166	12	7
11	129	7	5
12	115	4	3
13	84	1	1
Highest	258	33	13

Front Runners Impact Value = 2.6
82 winners from 258 races (32%)

5 FURLONGS 216 YARDS

Draw Pos	Winners	%
Low	123	33
Middle	133	36
High	114	31

DRAW NO	RUNNERS	WINNERS	%
1	354	33	9
2	344	46	13
3	353	39	11
4	356	38	11
5	351	30	9
6	350	40	11
7	334	25	7
8	316	37	12
9	293	25	9
10	265	21	8
11	223	12	5
12	208	13	6
13	167	11	7
Highest	370	43	12

Front Runners Impact Value = 1.8
81 winners from 370 races (22%)

7 FURLONGS 32 YARDS

Draw Pos	Winners	%
Low	150	36
Middle	158	38
High	106	26

DRAW NO	RUNNERS	WINNERS	%
1	392	52	13
2	387	47	12
3	393	49	12
4	401	45	11
5	388	40	10
6	383	42	11
7	375	38	10
8	348	32	9
9	336	23	7
10	312	22	7
11	274	16	6
12	231	8	3
Highest	414	44	11

Front Runners Impact Value = 1.7
92 winners from 414 races (22%)

1 MILE 141 YARDS

Draw Pos	Winners	%
Low	157	35
Middle	166	37
High	123	28

DRAW NO	RUNNERS	WINNERS	%
1	413	53	13
2	418	66	16
3	420	41	10
4	423	44	10
5	418	48	11
6	412	34	8
7	389	36	9
8	359	32	9
9	317	17	5
10	289	28	10
11	249	18	7
12	218	14	6
13	189	15	8
Highest	446	69	15

Front Runners Impact Value = 1.3
69 winners from 446 races (15%)

1 MILE 1 FURLONG 103 YARDS

Draw Pos	Winners	%
Low	91	29
Middle	128	41
High	94	30

DRAW NO	RUNNERS	WINNERS	%
1	299	27	9
2	300	30	10
3	292	41	14
4	298	28	9
5	303	32	11
6	295	27	9
7	277	31	11
8	261	25	10
9	239	22	9
10	210	16	8
11	179	15	8
12	156	12	8
13	135	7	5
Highest	313	42	13

Front Runners Impact Value = 1.1
40 winners from 313 races (13%)

1 MILE 4 FURLONGS 50 YARDS

Draw Pos	Winners	%
Low	82	35
Middle	93	39
High	62	26

DRAW NO	RUNNERS	WINNERS	%
1	225	34	15
2	226	32	14
3	224	18	8
4	222	28	13
5	222	24	11
6	221	24	11
7	207	20	10
8	190	22	12
9	175	14	8
10	149	7	5
11	125	7	6
12	105	7	7
Highest	237	25	11

Front Runners Impact Value = 1.6
45 winners from 237 races (19%)

1 MILE 5 FURLONGS 194 YARDS

Draw Pos		%
Low	27	28
Middle	44	45
High	27	28

DRAW NO	RUNNERS	WINNERS	%
1	91	6	7
2	96	7	7
3	92	12	13
4	91	17	19
5	91	8	9
6	94	14	15
7	87	8	9
8	81	6	7
9	72	5	7
10	66	5	8
11	62	4	6
12	55	4	7
13	45	2	4
Highest	98	12	12

Front Runners Impact Value = 1.1
13 winners from 98 races (13%)

2 MILES 119 YARDS

Draw Pos	Winners	%
Low	11	22
Middle	23	45
High	17	33

DRAW NO	RUNNERS	WINNERS	%
1	50	3	6
2	50	3	6
3	50	3	6
4	49	7	14
5	47	7	15
6	49	2	4
7	45	4	9
8	44	7	16
9	44	6	14
10	41	2	5
11	39	3	8
12	33	2	6
13	26	2	8
Highest	51	4	8

Front Runners Impact Value = 0.4
2 winners from 51 races (4%)

With a short run to the first bend in races over five furlongs, seven furlongs and a mile, low draws hold sway over these trips at Wolverhampton. The trend is also towards low draws over six furlongs, but with a straight run of well over a quarter of a mile to the bend out of the back straight, the bias is less pronounced. Early pace is also crucial, especially over the extended five furlongs where front-runners win nearly two-and-a-half times more often than they are entitled to.

There is little in the way of a bias over most of the longer trips, though interestingly a low-to-middle draw and early pace seems to be an asset over a mile and a half.

KEMPTON

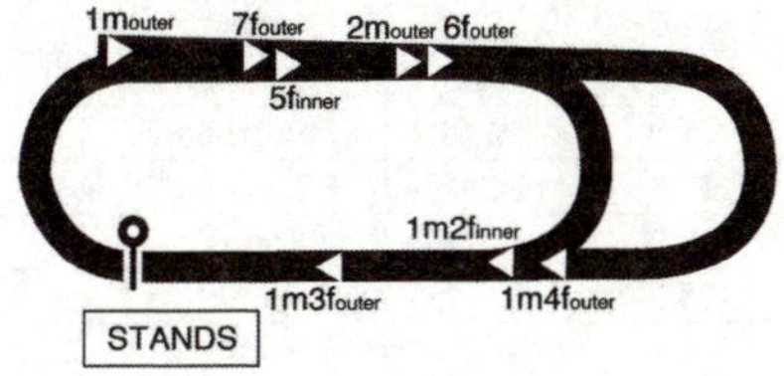

KEMPTON was obviously a well-established racecourse, staging decent Flat racing and high-class jumping, even before they built their Polytrack circuit, which came into being in March 2006. The powers that be rightly decided to retain the jumps track, which migrated outwards on to the old Flat turf course, but the Polytrack circuit follows a very different path.

In fact there are two All-Weather circuits; the main one is about ten furlongs round with a run-in of around three furlongs, whilst races run over five furlongs and a mile and a quarter use the inner bend (making the circuit just a mile round) with a run-in of about two furlongs. Where the two courses meet in the home straight the track widens for horses coming down the longer home straight and this 'cutaway' has been used to great effect by some jockeys. There are some interesting examples of this described in more detail later on in the book.

The All-Weather track also has the use of floodlights and after a quiet start the crowds for the track's themed Wednesday evening fixtures are starting to pick up. Hopefully that trend will continue and considering the facilities available to the public they certainly deserve decent patronage.

The circuit is unique in All-Weather racing in this country in that it is right-handed, and it soon became apparent that many horses preferred that orientation and therefore showed their best form here. That is certainly something well worth bearing in mind when analysing a Kempton All-Weather contest.

Another major asset for Kempton is that despite the change of surface, the track has retained its higher-quality established Flat races, with the Rosebery Handicap and two Group 3 contests (the September Stakes and the Sirenia Stakes for two-year-olds) still holding their places as major events in the Flat racing calendar. The quality of horse seen here is high in relative terms (the 2007 St Leger winner Lucarno broke his maiden here) and with several Listed races also being staged at the track during the course of the year, Kempton has done more than its fair share of raising the profile of All-Weather racing in this country.

Kempton Statistics March 2006 to July 2008

JOCKEYS

	TOTAL W-R	PER CENT	£1 LEVEL STAKE
Richard Hughes	43-228	19.0	+29.64
Dane O'Neill	39-352	11.0	-27.47
Jim Crowley	38-309	12.0	-21.78
George Baker	35-202	17.0	+51.85
Jamie Spencer	34-194	18.0	-59.76
Seb Sanders	31-262	12.0	-104.91
John Egan	30-204	15.0	+43.26
Ryan Moore	26-171	15.0	-37.77
Eddie Ahern	25-219	11.0	-61.56
Steve Drowne	24-288	8.0	-120.38
Fergus Sweeney	23-209	11.0	+10.29
Jimmy Fortune	22-168	13.0	-37.46
Martin Dwyer	22-146	15.0	+17.23
N Callan	22-182	12.0	-69.72
Ian Mongan	22-184	12.0	-28.98
Jimmy Quinn	20-221	9.0	-7.13
Micky Fenton	20-216	9.0	-34.05
Adam Kirby	20-220	9.0	-80.72
Chris Catlin	19-344	6.0	-136.66
T P Queally	19-220	9.0	-112.38
Oscar Urbina	18-139	13.0	+1.95

SIRES

	TOTAL W-R	PER CENT	£1 LEVEL STAKE
Dansili	24-169	14.2	+14.85
Cadeaux Genereux	21-114	18.4	+33.74
Fasliyev	21-138	15.2	+29.37
Danetime	20-129	15.5	+1.17
Bertolini	17-182	9.3	-90.93
Mark Of Esteem	17-154	11.0	-54.29
Cape Cross	17-130	13.1	-36.67
Compton Place	17-146	11.6	-24.31
Pivotal	17-116	14.7	-29.62
Piccolo	16-150	10.7	-31.38

FAVOURITES

	W-R	PER CENT	£1 LEVEL STAKE
2yo Stks	50-125	40.0	-4.89
3yo Stks	20-60	33.3	-11.08
3yo+ Stks	82-282	29.1	-50.32
Totals	**152-467**	**32.5**	**-66.29**
2yo H'caps	11-48	22.9	-8.58
3yo H'caps	47-152	30.9	-4.90
3yo+ H'caps	140-585	23.9	-60.33
Totals	**198-785**	**25.2**	**-73.81**
Apprentice	12-50	24.0	-8.66
Amateurs	0-5	0.0	-5.00
Totals	**12-55**	**21.8**	**-13.66**
All Favs	362-1307	27.7	-153.76

Kempton All-Weather Track

TRAINERS

	TOTAL W-R	2YO STKS	3YO STKS	OTHER STKS	2YO H'CPS	3YO H'CPS	OTHER H'CPS	APP	AMAT	PER CENT	£1 LEVEL STAKE
R Hannon	42-286	16-83	3-34	5-37	1-28	10-49	6-51	1-4	0-0	14.7	-89.77
G L Moore	35-205	1-6	0-0	10-49	0-0	0-8	24-131	0-9	0-2	17.1	-15.96
A M Balding	23-142	3-24	1-13	5-29	1-4	6-29	7-37	0-6	0-0	16.2	+32.38
D Shaw	22-200	1-5	0-1	2-26	0-11	2-12	17-141	0-4	0-0	11.0	-58.02
M Johnston	20-108	3-21	4-10	3-16	3-12	6-30	1-17	0-2	0-0	18.5	+17.06
A W Carroll	17-174	0-0	0-0	8-80	0-3	1-7	8-79	0-5	0-0	9.8	-24.75
J A Osborne	17-106	2-20	2-16	3-13	4-15	3-19	3-21	0-2	0-0	16.0	+6.59
A P Jarvis	16-155	2-16	2-9	3-32	1-7	4-31	4-60	0-0	0-0	10.3	-30.95
B W Hills	16-91	7-20	0-7	4-12	1-6	2-20	2-24	0-2	0-0	17.6	+12.16
M Botti	16-82	3-18	1-3	4-21	0-1	4-24	1-11	3-4	0-0	19.5	+54.88
P Howling	15-162	0-3	1-4	4-51	1-2	0-9	9-89	0-4	0-0	9.3	-42.25
D K Ivory	15-130	1-17	0-7	2-24	0-4	0-9	11-66	1-3	0-0	11.5	-18.25
J R Best	14-141	6-18	0-3	2-17	2-9	0-9	4-83	0-2	0-0	9.9	-52.14
P F I Cole	13-94	3-15	1-6	2-17	0-4	4-19	3-30	0-3	0-0	13.8	-13.63
M A Jarvis	13-75	2-15	3-9	5-14	0-1	2-16	1-19	0-1	0-0	17.3	-10.03
J H M Gosden	13-66	5-20	1-7	4-18	2-3	1-14	0-4	0-0	0-0	19.7	-6.33
B J Meehan	13-108	6-31	1-12	0-14	0-11	3-28	3-12	0-0	0-0	12.0	-42.46
J W Hills	12-107	0-15	0-6	0-13	0-7	2-19	7-35	3-12	0-0	11.2	-31.00
Sir M Stoute	12-47	2-8	1-5	3-14	1-1	2-9	1-8	2-2	0-0	25.5	+9.00
K R Burke	12-66	1-3	2-4	6-23	0-2	0-8	3-26	0-0	0-0	18.2	+37.53
S Kirk	12-93	6-29	2-7	0-18	1-8	2-18	0-12	1-1	0-0	12.9	-37.17

Kempton Draw and Front-Runner Statistics March 2006 to July 2008

5 FURLONGS

Draw Pos	Winners	%
Low	21	21
Middle	30	30
High	48	48

FROM RAIL	RUNNERS	WINNERS	%
1	96	19	20
2	99	10	10
3	91	16	18
4	94	11	12
5	89	2	2
6	84	3	4
7	78	9	12
8	61	6	10
9	57	7	12
10	45	5	11
11	37	0	0
12	25	1	4

Front Runners Impact Value = 2.8
33 winners from 99 races (33%)

6 FURLONGS

Draw Pos	Winners	%
Low	41	20
Middle	74	35
High	95	45

FROM RAIL	RUNNERS	WINNERS	%
1	201	26	13
2	198	32	16
3	193	23	12
4	197	20	10
5	200	17	9
6	190	16	8
7	181	16	9
8	170	13	8
9	148	8	5
10	115	4	3
11	88	9	10
12	52	5	10
13	2	0	0

Front Runners Impact Value = 1.8
41 winners from 210 races (20%)

7 FURLONGS

Draw Pos	Winners	%
Low	45	23
Middle	72	37
High	77	40

FROM RAIL	RUNNERS	WINNERS	%
1	183	30	16
2	185	19	10
3	182	14	8
4	189	16	8
5	188	25	13
6	182	16	9
7	175	12	7
8	162	18	11
9	152	7	5
10	146	9	6
11	126	7	6
12	112	4	4
13	81	7	9
14	48	3	6
15	1	0	0

Front Runners Impact Value = 1.8
34 winners from 194 races (18%)

1 MILE

Draw Pos	Winners	%
Low	68	26
Middle	104	40
High	89	34

FROM RAIL	RUNNERS	WINNERS	%
1	246	26	11
2	252	21	8
3	253	24	9
4	244	33	14
5	241	25	10
6	238	23	10
7	228	17	7
8	215	17	8
9	192	23	12
10	174	10	6
11	148	16	11
12	132	6	5
13	103	5	5
14	56	3	5
15	5	0	0
16	2	0	0

Front Runners Impact Value = 1.6
41 winners from 261 races (16%)

1 MILE 2 FURLONGS

Draw Pos	Winners	%
Low	35	28
Middle	51	40
High	40	32

FROM RAIL	RUNNERS	WINNERS	%
1	122	8	7
2	125	12	10
3	121	12	10
4	119	12	10
5	123	10	8
6	117	16	14
7	109	5	5
8	108	7	6
9	101	8	8
10	92	8	9
11	79	10	13
12	61	4	7
13	50	4	8
14	32	1	3
15	3	0	0
16	2	0	0

Front Runners Impact Value = 1.3
16 winners from 126 races (13%)

1 MILE 3 FURLONGS

Draw Pos	Winners	%
Low	21	28
Middle	32	43
High	22	29

FROM RAIL	RUNNERS	WINNERS	%
1	69	6	9
2	69	5	7
3	69	7	10
4	65	7	11
5	67	4	6
6	63	10	16
7	63	6	10
8	55	5	9
9	50	4	8
10	42	5	12
11	37	2	5
12	25	1	4
13	22	2	9
14	12	2	17
15	2	0	0

Front Runners Impact Value = 0.7
5 winners from 75 races (7%)

1 MILE 4 FURLONGS

Draw Pos	Winners	%
Low	32	30
Middle	42	40
High	31	30

FROM RAIL	RUNNERS	WINNERS	%
1	101	13	13
2	100	10	10
3	101	10	10
4	98	14	14
5	99	8	8
6	92	7	8
7	86	8	9
8	80	7	9
9	68	3	4
10	57	8	14
11	52	7	13
12	44	5	11
13	34	1	3
14	19	2	11

Front Runners Impact Value = 1.3
16 winners from 105 races (15%)

2 MILES

Draw Pos	Winners	%
Low	16	25
Middle	27	42
High	21	33

FROM RAIL	RUNNERS	WINNERS	%
1	61	7	11
2	63	4	6
3	61	5	8
4	58	9	16
5	61	4	7
6	57	4	7
7	58	10	17
8	53	4	8
9	48	4	8
10	43	2	5
11	35	4	11
12	30	2	7
13	22	2	9
14	10	0	0

Front Runners Impact Value = 0.3
2 winners from 64 races (3%)

Because Kempton is the only right-handed All-Weather track in the British Isles, analysing draw statistics properly is a problem if you use the real stall numbers. For instance, whilst stall seven might be next to the inside rail in a seven-horse field, the same stall might be eight-wide of the rail in another contest if there were 14 runners, and the path the respective horses will take is likely to be very different.

In order to overcome this, in his book *Betting For A Living (Aesculus Press 1992)* Nick Mordin advocates "turning the draw numbers on their head" so that the stall next to the inside rail becomes stall one, the next stall two, and so on. Hence in the draw statistics shown on the previous pages, the title above the stall numbers says "From Rail" rather than "Draw No". This method does have its drawbacks, especially with non-runners, which can distort the figures unless you can find a way around it, but even allowing for that margin for error the resulting data reveals some startling statistics.

For instance, they reveal just what a massive advantage is enjoyed by those horses berthed close to the inside rail over distances from five to seven furlongs, and over the minimum trip a combination of early speed and a high draw is huge plus (there are examples of this scenario in action later in the book).

Over some of the longer trips there is an advantage to those that race handily, but it is only slight.

Kandidate: winner of his first three starts on the Kempton Polytrack and seen here (nearside) winning the 2008 Magnolia Stakes

GREAT LEIGHS

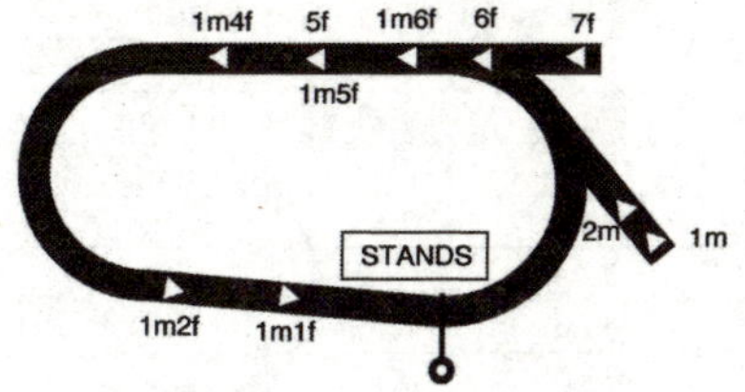

AFTER several frustrating delays, Great Leighs finally opened its doors on Sunday April 20th 2008 to rave reviews over the quality of the racing surface. The venue made its intentions clear right from the start with quality racing very much the name of the game, and therefore decent prize money has been put up. The track's executive have certainly been rewarded as some very decent horses have turned up here, which is all the more impressive when you consider that under BHA rules a brand new track cannot stage any contest above Class 2 level in its first year of operation. It will be fascinating to see what type of horses appear when that particular shackle is removed.

The surface at Great Leighs is Polytrack, the circuit is around nine-and-a-half furlongs in circumference and, as at Wolverhampton and Kempton, the track has the ability to race under floodlights. One of the first things which strikes you when you first see the circuit is just how wide the course is, and in some races the runners have tended to fan out a long way across the track on reaching the home straight. One or two horses have also hung all the way to the outside of the course rounding the home bend.

The bends themselves are wide and sweeping – the turn away from the stands (which are on the inside of the course) has a radius of 135 metres whilst the home bend is 142 metres – and each bend is around a quarter of a mile in length. The run-in from the home turn is just under two furlongs and on a track like this it would be assumed that it would suit galloping types. The ability to 'travel' seems to be of special importance here though, as the following statistic demonstrates.

Within the first 12 fixtures held between April and July 2008, a total of 12 horses were pulled up during the course of a race at Great Leighs, whilst there were just two in the same period at Southwell, one at Kempton, but none at either Lingfield or Wolverhampton. Admittedly of those 12 horses that were pulled up, one broke a blood vessel, one was pulled up with a slipping saddle and another virtually refused to race, but it still shows that there is little point in coming here with a horse that races lazily or cannot hold its position, because unless they can do so the track will punish them.

As far as a track bias is concerned and with betting-in-running in mind, it is well worth looking to see how horses fare if they stick tight to the inside rail up the home straight, because that has looked to be a disadvantage at some meetings.

Great Leighs Statistics April 2008 to July 2008

JOCKEYS

	TOTAL W-R	PER CENT	£1 LEVEL STAKE
Seb Sanders	5-28	18.0	-4.60
Jamie Spencer	4-24	17.0	-13.96
John Egan	3-21	14.0	-7.12
Jimmy Quinn	3-20	15.0	-5.75
Jimmy Fortune	3-8	38.0	+4.25
Robert Winston	3-10	30.0	+27.00
Pat Cosgrave	3-12	25.0	+30.75
Ryan Moore	3-10	30.0	0.00
Travis Block	3-7	43.0	+7.10
Michael Hills	2-3	67.0	-0.16
Philip Robinson	2-3	67.0	+5.25
James Doyle	2-14	14.0	-7.00
Darryll Holland	2-9	22.0	+5.50
Ted Durcan	2-17	12.0	-11.65
Fergal Lynch	2-5	40.0	+6.75
Eddie Ahern	2-13	15.0	-2.50
Jim Crowley	2-8	25.0	+2.83
George Baker	2-7	29.0	+1.25
Kerrin McEvoy	2-10	20.0	-4.00
Kirsty Milczarek	2-17	12.0	+5.50
Jack Mitchell	2-7	29.0	+36.00

SIRES

	TOTAL W-R	PER CENT	£1 LEVEL STAKE
Green Desert	4-12	33.3	+28.75
Oasis Dream	3-13	23.1	-0.50
Dansili	3-15	20.0	-3.75
Danehill	2-6	33.3	+15.00
Dr Fong	2-13	15.4	+3.00
Pivotal	2-9	22.2	-3.00
Catcher In The Rye	2-7	28.6	+7.50

FAVOURITES

	W-R	PER CENT	£1 LEVEL STAKE
2yo Stks	2-5	40.0	+0.15
3yo Stks	3-4	75.0	+2.46
3yo+ Stks	6-17	35.3	+1.04
Totals	**11-26**	**42.3**	**+3.65**
2yo H'caps	0-0	0.0	0.00
3yo H'caps	6-16	37.5	+1.58
3yo+ H'caps	13-40	32.5	+3.63
Totals	**19-56**	**33.9**	**+5.21**
Apprentice	0-2	0.0	-2.00
Amateurs	0-0	0.0	-0.00
Totals	**0-2**	**0.0**	**-2.00**
All Favs	30-84	35.7	+6.86

TRAINERS

	TOTAL W-R	2YO STKS	3YO STKS	OTHER STKS	2YO H'CPS	3YO H'CPS	OTHER H'CPS	APP	AMAT	PER CENT	£1 LEVEL STAKE
A J McCabe	5-22	0-2	0-1	0-1	0-0	3-5	2-13	0-0	0-0	22.7	+10.50
Sir M Prescott	3-8	0-1	0-0	0-2	0-0	0-1	3-4	0-0	0-0	37.5	+3.40
H R A Cecil	3-8	0-0	0-0	2-5	0-0	1-3	0-0	0-0	0-0	37.5	+0.60
M Johnston	3-13	0-0	0-0	0-4	0-0	3-4	0-5	0-0	0-0	23.1	+10.50
G L Moore	3-8	0-0	0-0	0-0	0-0	0-0	3-8	0-0	0-0	37.5	+4.25
L M Cumani	2-7	0-1	0-0	0-0	0-0	1-3	1-3	0-0	0-0	28.6	-3.37
J W Hills	2-7	0-0	0-0	1-2	0-0	0-1	1-4	0-0	0-0	28.6	+11.50
B W Hills	2-6	1-2	0-0	1-1	0-0	0-1	0-2	0-0	0-0	33.3	-3.16
M A Jarvis	2-4	0-0	1-1	0-0	0-0	1-2	0-1	0-0	0-0	50.0	+4.25
W J Haggas	2-8	0-0	1-1	1-3	0-0	0-3	0-1	0-0	0-0	25.0	-3.33
P D Evans	2-8	1-1	1-2	0-0	0-0	0-1	0-4	0-0	0-0	25.0	+15.87
R Ingram	2-6	0-0	0-0	0-0	0-0	1-3	1-3	0-0	0-0	33.3	+6.75
Gay Kelleway	2-13	0-0	0-1	1-3	0-0	0-2	1-7	0-0	0-0	15.4	-5.00
M Botti	2-12	0-2	0-2	2-3	0-0	0-4	0-1	0-0	0-0	16.7	-4.00
E A L Dunlop	2-12	0-2	0-0	2-4	0-0	0-5	0-1	0-0	0-0	16.7	-5.75
H Morrison	2-5	0-0	0-1	0-0	0-0	0-0	2-3	0-1	0-0	40.0	+1.60
M J Wallace	2-10	1-2	0-0	0-2	0-0	1-2	0-4	0-0	0-0	20.0	-2.50
W Jarvis	1-10	0-0	0-0	1-3	0-0	0-2	0-5	0-0	0-0	10.0	+5.00
C F Wall	1-6	0-0	0-0	0-1	0-0	0-0	1-5	0-0	0-0	16.7	+11.00
D Elsworth	1-12	0-0	0-0	1-8	0-0	0-3	0-1	0-0	0-0	8.3	-3.00
J D Bethell	1-1	0-0	0-0	0-0	0-0	0-0	1-1	0-0	0-0	100.0	+5.00

Great Leighs Draw and Front-Runner Statistics April 2008 to July 2008

5 FURLONGS

Draw Pos	Winners	%
Low	1	14
Middle	4	57
High	2	29

DRAW NO	RUNNERS	WINNERS	%
1	6	0	0
2	7	1	14
3	7	1	14
4	7	1	14
5	7	1	14
6	7	0	0
7	5	2	40
8	4	0	0
9	3	1	33
10	2	0	0
12	1	0	0
Highest	7	2	29

Front Runners Impact Value = 3.4
3 winners from 7 races (43%)

6 FURLONGS

Draw Pos	Winners	%
Low	3	17
Middle	7	39
High	8	44

DRAW NO	RUNNERS	WINNERS	%
1	17	2	12
2	15	1	7
3	16	0	0
4	17	3	18
5	17	3	18
6	18	1	6
7	13	2	15
8	13	2	15
9	9	1	11
10	8	1	13
11	3	0	0
12	3	1	33
13	2	0	0
14	2	0	0
15	2	1	50
16	2	0	0
Highest	18	4	22

Front Runners Impact Value = 1.0
2 winners from 18 races (11%)

1 MILE

Draw Pos	Winners	%
Low	6	38
Middle	3	19
High	7	44

DRAW NO	RUNNERS	WINNERS	%
1	15	1	7
2	16	1	6
3	15	0	0
4	15	5	33
5	15	1	7
6	15	1	7
7	13	2	15
8	13	0	0
9	10	2	20
10	7	1	14
11	9	1	11
12	7	0	0
13	6	0	0
14	5	0	0
15	6	1	17
16	6	0	0
Highest	16	3	19

Front Runners Impact Value = 1.3
2 winners from 16 races (13%)

1 MILE 2 FURLONGS

Draw Pos	Winners	%
Low	2	11
Middle	9	47
High	8	42

DRAW NO	RUNNERS	WINNERS	%
1	18	1	6
2	18	1	6
3	18	3	17
4	18	3	17
5	18	2	11
6	17	0	0
7	16	3	19
8	12	2	17
9	8	0	0
10	9	1	11
11	6	0	0
12	4	1	25
13	4	1	25
14	3	0	0
15	3	1	33
16	2	0	0
Highest	19	4	21

Front Runners Impact Value = 0.8
2 winners from 19 races (11%)

1 MILE 6 FURLONGS

Draw Pos	Winners	%
Low	1	14
Middle	4	57
High	2	29

DRAW NO	RUNNERS	WINNERS	%
1	7	0	0
2	7	1	14
3	6	0	0
4	6	1	17
5	7	2	29
6	8	0	0
7	6	2	33
8	3	0	0
9	3	1	33
10	2	0	0
11	2	0	0
12	1	0	0
13	1	0	0
14	1	0	0
15	1	0	0
Highest	7	1	14

Front Runners Impact Value = 0.0
0 winners from 7 races (0%)

2 MILES

Draw Pos	Winners	%
Low	1	50
Middle	0	0
High	1	50

Draw No	Runners	Winners	%
1	2	0	0
2	2	0	0
3	2	0	0
4	2	1	50
5	2	0	0
6	2	1	50
7	1	0	0
8	1	0	0
9	1	0	0
10	1	0	0
11	1	0	0
12	1	0	0
13	1	0	0
14	1	0	0
15	1	0	0
Highest	2	1	50

Front Runners Impact Value = 5.3
1 winner from 2 races (50%)

With Great Leighs still very much in its infancy, we can't make too many hard and fast conclusions about the effects of the draw and it may not be until 2009 or 2010 that we can begin to make proper judgements. The intention when the track was being built was that they would try and make sure there was no significant draw bias and at this stage there is little evidence to suggest that they have failed in that venture. More evidence will be needed to see whether they have truly succeeded.

Front-runners have done well in the few races over five furlongs so far, which is the same as for most of the other All-Weather tracks, but again it is still early days.

Great Leighs All-Weather Track

DUNDALK

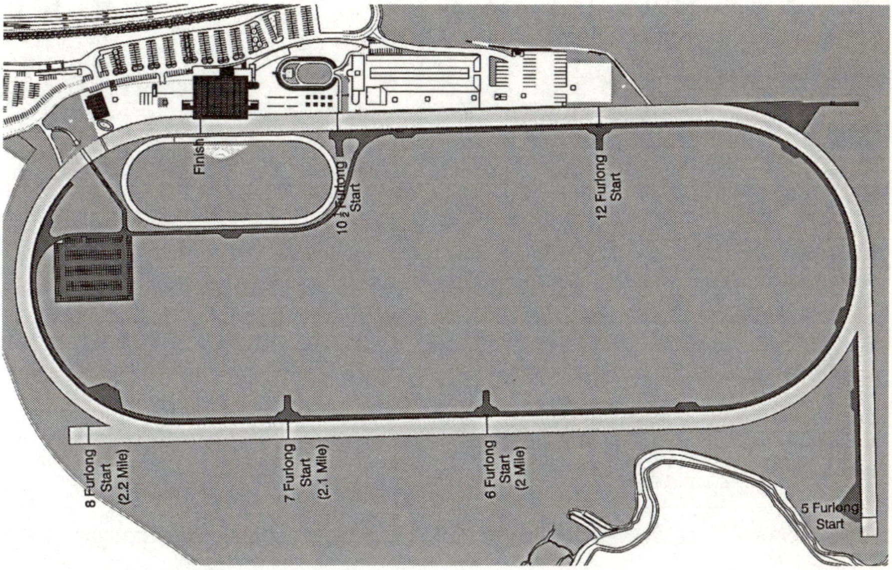

LEGEND has it that horse racing started on the beach at Laytown in 1876, though the racing as we would recognise it today started a bit later in 1901. Therefore Laytown can claim to be the oldest currently operational sand track in the British Isles, and by some way, but Dundalk became the first track in Ireland to stage a fixture on an artificial surface in August 2007.

The majority of the meetings take place under floodlights on a Friday evening and as with Great Leighs, those in charge at Dundalk seemed determined to make sure that the racing was of sufficiently high quality from the very start. The top Irish trainers were soon represented there and even some British stables have sent horses over, attracted by the decent prize money on offer. Certainly the track has drawn praise from top jockeys and trainers alike and at this stage the future looks very bright for the track in the northeast corner of the Irish Republic.

The circuit itself is Polytrack and is a left-handed oval of about ten furlongs, with a separate chute for races over five furlongs which joins the home straight at an angle – not unlike the sprint track at Warwick – but with a rather longer run-in from the home turn of around three furlongs.

Dundalk Statistics August 2007 to July 2008

JOCKEYS

	TOTAL W-R	PER CENT	£1 LEVEL STAKE
F M Berry	21-123	17.1	+10.12
J Murtagh	20-62	32.3	+52.23
D P McDonogh	16-115	13.9	-19.25
P J Smullen	14-117	12.0	-18.50
J A Heffernan	11-112	9.8	-13.67
K J Manning	10-89	11.2	+5.50
M J Kinane	9-41	22.0	-0.68
C D Hayes	9-97	9.3	-25.88
W M Lordan	7-86	8.1	-37.50
C O'Donoghue	6-37	16.2	+33.00
E J McNamara	6-43	14.0	+3.35
R P Cleary	5-71	7.0	+13.50
W J Lee	5-88	5.7	-34.50
W J Supple	4-70	5.7	-38.50
K Latham	3-17	17.6	-9.31
D M Grant	3-32	9.4	-14.50

SIRES

	TOTAL W-R	PER CENT	£1 LEVEL STAKE
Val Royal	6-18	33.3	+20.85
Danetime	6-34	17.6	+2.84
Galileo	5-23	21.7	+6.82
Elnadim	5-40	12.5	-12.65
Invincible Spirit	5-52	9.6	-24.25
Danehill	5-31	16.1	-8.50
Danehill Dancer	5-26	19.2	+9.33
Mujadil	4-20	20.0	+22.00
Stravinsky	4-10	40.0	+29.75
Rock Of Gibraltar	4-17	23.5	+12.50
Noverre	4-19	21.1	+19.50

FAVOURITES

	W-R	PER CENT	£1 LEVEL STAKE
2yo Stks	9-32	28.1	-10.80
3yo Stks	6-24	25.0	-7.06
3yo+ Stks	10-31	32.3	-4.26
Totals	**25-87**	**28.7**	**-22.12**
2yo H'caps	2-9	22.2	-2.00
3yo H'caps	6-21	28.6	+2.00
3yo+ H'caps	18-81	22.2	-1.90
Totals	**26-111**	**23.4**	**-1.90**
Apprentice	2-6	33.3	+2.00
Amateurs	0-0	0.0	-0.00
Totals	**2-6**	**33.3**	**+2.00**
All Favs	53-204	26.0	-22.02

Dundalk All-Weather Track

TRAINERS

	TOTAL W-R	2YO STKS	3YO STKS	OTHER STKS	2YO H'CPS	3YO H'CPS	OTHER H'CPS	APP	AMAT	PER CENT	£1 LEVEL STAKE
G M Lyons	19-104	5-24	4-17	3-13	0-5	2-12	5-31	0-2	0-0	18.3	+3.37
John M Oxx	12-42	3-8	4-11	3-13	0-2	2-5	0-3	0-0	0-0	28.6	-2.56
D K Weld	10-68	2-13	2-17	3-15	1-5	2-7	0-11	0-0	0-0	14.7	-6.50
M J P O'Brien	7-19	0-0	0-1	4-8	0-0	0-0	3-10	0-0	0-0	36.8	+50.00
P D Deegan	7-34	1-9	1-4	0-1	1-1	3-10	1-8	0-1	0-0	20.6	+33.50
T Stack	6-15	1-2	1-3	0-3	0-0	0-0	4-7	0-0	0-0	40.0	+35.23
M Halford	6-69	1-18	1-7	1-6	0-1	0-7	3-28	0-2	0-0	8.7	+1.50
K Prendergast	6-54	2-19	1-12	3-8	0-6	0-4	0-5	0-0	0-0	11.1	-29.88
J S Bolger	5-43	2-16	0-5	0-3	0-2	1-8	2-9	0-0	0-0	11.6	-9.00
E Lynam	5-43	1-9	1-12	0-8	1-3	1-3	1-8	0-0	0-0	11.6	-7.25
C O'Brien	5-29	1-4	0-3	0-1	1-3	0-2	3-15	0-1	0-0	17.2	-3.50
Eoin Griffin	5-34	0-7	0-3	1-6	0-2	2-3	2-12	0-1	0-0	14.7	-2.20
Patrick Flynn	4-12	0-0	1-1	0-2	0-0	0-0	3-9	0-0	0-0	33.3	+9.50
A P O'Brien	4-29	3-15	0-5	1-7	0-2	0-0	0-0	0-0	0-0	13.8	-7.00
T G McCourt	4-38	0-1	0-0	0-5	0-0	0-1	3-28	1-3	0-0	10.5	+10.50
Noel Meade	4-12	0-0	1-2	1-3	0-0	0-0	1-6	1-1	0-0	33.3	+7.50
Eoin Doyle	4-14	0-0	0-0	0-2	0-0	0-0	4-12	0-0	0-0	28.6	+3.35
Patrick Martin	3-41	0-8	0-7	1-2	0-1	2-5	0-17	0-1	0-0	7.3	-26.75
Ms F Crowley	3-37	0-5	0-9	1-5	0-2	0-4	2-9	0-3	0-0	8.1	-14.50
A McGuinness	3-37	0-0	0-1	0-3	0-0	0-2	3-30	0-1	0-0	8.1	-23.00
Liam Roche	3-24	1-7	1-5	0-1	1-5	0-3	0-2	0-1	0-0	12.5	+7.50

Dundalk Draw and Front-Runner Statistics August 2007 to July 2008

5 FURLONGS

Draw Pos	Winners	%
Low	10	56
Middle	3	17
High	5	28

DRAW NO	RUNNERS	WINNERS	%
1	18	2	11
2	16	5	31
3	17	3	18
4	18	1	6
5	17	1	6
6	18	2	11
7	14	1	7
8	16	1	6
9	15	0	0
10	13	2	15
11	11	0	0
12	11	0	0
13	11	0	0
14	8	0	0
Highest	18	3	17

Front Runners Impact Value = 1.4
3 winners from 18 races (17%)

6 FURLONGS

Draw Pos	Winners	%
Low	10	36
Middle	9	32
High	9	32

DRAW NO	RUNNERS	WINNERS	%
1	26	0	0
2	27	1	4
3	27	4	15
4	26	7	27
5	26	2	8
6	26	3	12
7	27	2	7
8	26	2	8
9	26	2	8
10	23	1	4
11	22	1	5
12	19	1	5
13	17	0	0
14	16	2	13
Highest	28	6	21

Front Runners Impact Value = 0.3
1 winner from 28 races (4%)

7 FURLONGS

Draw Pos	Winners	%
Low	10	34
Middle	12	41
High	7	24

DRAW NO	RUNNERS	WINNERS	%
1	29	4	14
2	29	0	0
3	29	5	17
4	29	5	17
5	29	3	10
6	28	1	4
7	27	2	7
8	25	3	12
9	23	2	9
10	22	0	0
11	21	1	5
12	20	2	10
13	20	1	5
14	18	0	0
Highest	29	2	7

Front Runners Impact Value = 1.6
5 winners from 29 races (17%)

1 MILE

Draw Pos	Winners	%
Low	14	
Middle	23	48
High	11	23

Draw No	Runners	Winners	%
1	48	5	10
2	48	2	4
3	48	5	10
4	45	5	11
5	47	8	17
6	44	6	14
7	45	2	4
8	44	3	7
9	41	3	7
10	38	3	8
11	39	4	10
12	37	1	3
13	37	1	3
14	31	0	0
15	1	0	0
Highest	48	1	2

Front Runners Impact Value = 0.6
3 winners from 48 races (6%)

1 MILE 2 FURLONGS 150 YARDS

Draw Pos	Winners	%
Low	8	23
Middle	16	46
High	11	31

DRAW NO	RUNNERS	WINNERS	%
1	34	2	6
2	34	2	6
3	35	3	9
4	35	1	3
5	35	2	6
6	35	4	11
7	35	6	17
8	31	4	13
9	30	4	13
10	29	2	7
11	27	2	7
12	24	1	4
13	24	1	4
14	21	1	5
Highest	35	4	11

Front Runners Impact Value = 0.3
1 winner from 35 races (3%)

1 MILE 4 FURLONGS

Draw Pos	Winners	%
Low	3	27
Middle	4	36
High	4	36

DRAW NO	RUNNERS	WINNERS	%
1	11	1	9
2	11	0	0
3	11	1	9
4	11	1	9
5	11	2	18
6	10	0	0
7	10	2	20
8	10	0	0
9	9	0	0
10	9	0	0
11	9	3	33
12	9	0	0
13	8	0	0
14	8	1	13
15	1	0	0
Highest	11	0	0

Front Runners Impact Value = 1.3
1 winner from 11 races (9%)

It's still relatively early days at Dundalk and with such limited evidence to go on there aren't too many conclusions we can make about the effects of the draw or the suitability to front-runners.

At this stage low-to-middle draws seem to hold the advantage at up to a mile and low draws do especially well over five furlongs, but the statistics for front-runners seem very random and we need to wait a bit longer for any pattern to truly emerge.

USING RATINGS

PEOPLE who bet on horses, whether regular punters or not, use all sorts of methods to come up with their selections. At one end of the scale are the academics and the scientists, plus those that are just very talented or very shrewd, and through their hard endeavour they have found ways to make their betting pay. At the other end of the spectrum are those who back horses because the saddlecloth number happens to be their lucky number, or the jockey's silks look nice, or the horse's name contains some sort of personal relevance (the horse is called Flash Harry and my favourite uncle is called Harry, etc, etc).

Before anyone accuses me of being sexist and that the second group will probably contain a greater number of females than the first, then they are partly right, but I'm not being sexist – my wife would never allow it. For example, I was at Lingfield on a cold wet day at the beginning of 2008 when I overheard a lady say to her friends: "Oh look, there's a horse in this race called Happy As Larry. I've got to back him haven't I? Because that's me, isn't it ... Happy As Larry?"

Her friends all smiled the sort of smile you come up with when you're trying to humour someone. Happy As Larry came in fifth.

The fact of the matter is that betting on horses, especially to decent amounts, remains a male-orientated pursuit more's the pity. However, the maddening thing for those of us that take our betting seriously is that the small-time punters, whose methods we often snigger at, occasionally produce a very big winner and when it happens we certainly get to know about it. The publicity machine quickly grinds into action and to be honest can you blame them? After all, the "it could be you" message will strike more of a resonance with the occasional and/or small-stake punter's mentality and show them that you don't need to be an expert or put in much work in order to win a huge sum of money.

Serious punters don't need any encouragement to keep on betting, hence it's better for the bookmakers to focus on the occasional punter who manages to nail a massive accumulator because their last selection had the same name as next door's cat, rather than a boring old fart that often stays up half the night surrounded by form books. The problem is that for every successful small-stake punter, there are thousands who rarely win any substantial amount of money, though most of them aren't that bothered because it's just fun!

So which tools are the best for making money at this All-Weather game? In my opinion some sort of performance ratings are a massive aid in the whole selection process and, whilst they are far from the be-all and end-all, without them it's hard to get an initial grip on any contest, at least it is for me. However, once I have a set of ratings in front of me that I feel comfortable with, I can quickly start to dismiss the no-hopers and in doing so can reduce the number of horses that I need to take a closer look at.

The question you need to ask is which type of ratings do you wish to use. My own experience with ratings stretches back to when I first became interested in horse racing in the late 1960s. In those days my late father was an avid reader of the *Sporting Life*, so my first experience of ratings were those of the late Dick Whitford and since then I've been through many phases, some of them lasting several years.

Amongst the other ratings services I have used in my lifetime are those of *Timeform*, *Raceform*, the old *Haig Annual*, and even Formcast in the *Daily Mail*. I can put my hand on my heart and say that I enjoyed some very big wins thanks to all of them and I can't really say that I considered any of these products to have been much better or worse than any of the others. I went through long winning runs with all of them, and also long losing runs with all of them. What I tended to forget though, and this is a problem with any widely available ratings service, is that if I liked the look of a particular horse based on the publisher's ratings, then so would thousands of others. I used to think that I was in a special position because I knew something that very few people did, but of course that was completely wrong.

For me, the choice of which type of ratings I wanted to specialise in became obvious in the early 1990s. Having become the latest incarnation of *Split Second* in 2000, of course I'm biased towards speed figures, but I had done plenty of experimentation with them prior to that before I decided that this was the way I wanted to go.

It was around 1992 that I became very interested in speed figures. As I have already said, before then I was using various types of performance ratings and fair play to those people who compile them, because they are totally opinion-based or subjective with no particular 'constant' to support them. It's a bit like a tightrope walker trying to get from one end to the other without the aid of a balancing pole. In the case of speed ratings, the 'pole' is the time taken for the race to be run and that also provides the 'constant' referred to earlier.

As I have said many times before, a race time cannot be quibbled with provided it has been measured accurately. If a five-furlong sprint at Southwell takes 59 seconds to be run, then 59 seconds will go down in the official formbook and other various publications. It can't be twisted one way or the other, unless of course the timing equipment is faulty or the race hasn't been run over the advertised distance. The only question then is how to interpret it.

I should say at this point that most of what I am about to describe can be applied to turf racing (Flat and Jumps) and indeed that is what I have to do as *Split Second*, but All-Weather racing is something different and it's all to do with pace and tempo. As we all know, even these days many turf races are run a different way to those on sand. Even though the differences are not quite as stark as they used to be in the early days of All-Weather racing when the horses used to go hell for leather straight from the stalls and it was almost a case of last man standing, tempo is still all-important. Admittedly the introduction of Polytrack has made it more possible to ride a waiting race, especially over the longer distances at Lingfield thanks mainly to the layout of the track, but even so in most cases it still pays to race handily and that means that the majority of races on sand are truly run. Hence the value of speed ratings.

My first real venture into betting on the All-Weather followed soon after my conversion to the merits of speed ratings, and it wasn't long before these two fitted together hand in glove. My first visit to an All-Weather meeting came in January 1994 when due to the turf fixtures being snowed off I ended up enjoying a cold Friday at Southwell. Things soon warmed up though thanks to a string of winners courtesy of my hastily calculated and rather amateurish speed ratings. I ended up winning a tidy sum and felt that if I could do that with a set of cobbled-together ratings, what could I achieve if I did it properly? The following year I came up with a method that I was comfortable with and went about calculating some standard times and a formula to calculate the actual ratings themselves. Fortunately I was able to write some rather amateurish computer programs to make the whole thing a bit more manageable and less time-consuming.

These are the methods I originally used in order to establish my initial benchmarks, followed by the way I calculate my own speed figures to this day. I'm afraid it may all seem a bit involved, complex and time consuming at first, but I'm very happy with the results of my labours and some of what I describe here only ever needs to be done once. There are plenty of people around who compile their own speed ratings and use very different methods, but because my speed ratings appear regularly throughout this book it seems only right that I should show everything that has gone into bringing them about.

The first thing that has to be done is to establish a set of times next to which you can measure the performance of any given winner. These times, known generally as standard times, must be calculated the same way for every distance used at each All-Weather track. Although the number of sand tracks in the British Isles has doubled within the past three years, the task is still a lot easier than having to do it for every turf track.

The way I went about this was to record the winning times of recent races over every distance at each All-Weather track, going back far enough to provide a large enough sample. I originally set the minimum number of sample races for each distance to 11 (the reason for using an odd number will become evident), but because I limited my search so that races confined to younger horses or fillies were omitted, I found myself having to go back further than I really wanted to and therefore having to delve into more results annuals than ideal. Many races that are open to both sexes are won by fillies anyway, and indeed many races open to older horses are won by three-year-olds, so I don't now really see the need to omit any race except those restricted to two-year-olds, because they are definitely run in significantly slower times. If they were included they would distort the final sample.

Once you have been through this process, you would end up with a list of 11 winning times for each distance at each track (you can of course decide on whatever minimum sample you want, but the larger the sample the more accurate and reliable your standard times will be). Once completed, these listings should be sorted into ascending order, so that for the following three distances at a hypothetical track you might have the following –

6f	**7f**	**1m**
1m 10.54	1m 25.02	1m 38.67
1m 11.65	1m 25.11	1m 38.77
1m 11.74	1m 25.35	1m 38.82
1m 11.81	1m 25.47	1m 39.21
1m 12.62	1m 26.01	1m 39.78
1m 13.18 *	1m 26.11 *	1m 40.09 *
1m 13.85	1m 26.34	1m 40.30
1m 13.90	1m 27.43	1m 40.44
1m 14.04	1m 27.80	1m 40.82
1m 14.55	1m 27.83	1m 41.29
1m 15.97	1m 28.22	1m 41.56

Notice the times next to the asterisks, the reason they are where they are is that they indicate the sixth items on three 11-item lists. Had the samples been 21-item lists, then the arrows would have been pointing at the 11th item. Either way they are pointing at the middle item (hence the reason for having an odd number of race samples). This method means that you will end up with a true average time taken to cover the respective race distances within the sample period. Notice also that over six furlongs the fastest and slowest times are well adrift of the nearest times to them, perhaps freakishly so, but this method means that they will not distort the results.

Note: As these 'standard' times are the middle items in the list, they are sometimes referred to as median times.

Having established the median times, we now need to establish a set of benchmarks to help measure the merit of each winning performance against the clock. It will also help establish how much the conditions (i.e. the state of the track, wind, rail movement, etc) have affected the winning times. In this book this effect is called the "track variant", though in other places it may be referred to as the "going allowance" or the "going correction". The idea is that classier horses will run faster times relative to modest horses and we need to find out what that 'class factor' is. This is the way I went about it.

Again I looked back at recent results and compared each winning time with the median time for that course and distance. For instance, using the example median times shown above, had the race been run over six furlongs and the winning time had been 1m 12.78, then the time would have been 0.4 seconds fast (the median time is 1m 13.18). It's not enough to write this difference down just like that, however, as another calculation needs to be made.

It's probably fairly obvious that the longer the race distance, the greater the potential margin between the winning time and the median time is likely to be. To use an extreme example, I can get within four seconds of the 100 metres world record on an athletics track even though I am not very fit, but if I got within four seconds of the marathon world record, then I would be a gold-medal candidate. So as I see it, if a horse beats the median time by 0.5 of a second over five furlongs and another beats the median time by 1.6 seconds over two miles (or 16 furlongs), they have put up the same performance provided the going was consistent. In other words they have run $1/10^{th}$ of a second faster than the median time for every furlong they ran. This means that the original difference of 0.4 seconds fast needs to be adjusted so that all winning performances can be measured against each other, regardless of the trip. The way I do this is that I assume that all races have been run over one mile and adjust the difference between the winning time and the median time accordingly.

In our example, the winner ran faster than the median time for six furlongs by 0.4 seconds. Had he maintained the same velocity up to one mile, then he would have beaten the median time by 0.53 seconds (8/6 x 4) and it is that time (-0.53) that is recorded. If the race had been over one mile and six furlongs and the winning time had been 1.8 seconds slower than the median time, then the adjusted difference would be +1.03 (8/14 x 1.8).

How we record these adjusted differences is very important. What I did back in 1995 was to write all the race classes across the top of a sheet of paper from left to right. In those days the classes were listed from A to G, but these days the classes range from 1 to 7 on the Flat. Let's say that our example six-furlong race was a Class 3 contest, whilst the one-mile-six-furlong race was a Class 5, then –0.53 would go in the Class 3 column and +1.03 would go under Class 5. Remember that it doesn't matter where the race was or over what distance, once the necessary adjustments have been made to the margin between the winning time and median time as shown above, the important thing is to put the right item in the correct "class" column.

You need a large enough sample of adjusted differences (again ignoring races confined to two-year-olds) to give reliable results. In 1995 I made the minimum number of sample races 20, which meant I had to go back a fair way for some of the lesser-used distances, but this was one of the tasks I was only going to have to do once. What I was trying to do was establish a set of class pars.

Note: in 1995 there was only one Class A event run on sand – the Wulfrun Stakes at Wolverhampton – so that particular class was ignored and I just concentrated on Classes B to F (or 2 to 6).

Once I had a large enough sample for each race class, I once again sorted the items within each list so that the fastest winning times compared to median were at the top and the slowest at the bottom. For example, the items within the Class 4 and Class 5 listings looked something like this –

	Class 4		Class 5
*	-1.42	*	-1.12
*	-1.37	*	-0.89
*	-1.25	*	-0.75
*	-0.81	*	-0.54
*	-0.75	*	-0.33
*	-0.64	*	-0.12
	-0.51		-0.05
	-0.43		-0.01
	-0.32		+0.07
	-0.17		+0.12
	+0.11		+0.38
	+0.17		+0.51
	+0.29		+0.69
	+0.45		+0.78
	+0.58		+0.93
	+0.78		+1.13
	+0.98		+1.27
	+1.10		+1.41
	+1.59		+1.85
	+1.91		+2.05

The way I decided to establish the class pars was to average the better performances within each listing. I therefore took the top six adjusted differences from each list (shown next to the asterisks) and averaged them. The result for the Class 4 category was -1.04 and for Class 5 it was -0.625. What does this tell us? Well if we were to allocate a speed rating to each of these averages then the following would happen.

On my speed ratings scale, one point equals one length per mile and on the Flat one length is considered to be about 1/5 (or 0.2) of a second. The average of the better performances in the Class 4 category was -1.04 seconds, which is 5.2 lengths (-1.04 / 0.2). That means that the better Class 4 winners had beaten the median time by an average of 5.2 lengths per mile. The average for the better winners within the Class 5 category was –0.625. Dividing that by 0.2 gives -3.125, which means that the better Class 5 winners beat the median time by an average of 3.125 lengths per mile. In fact what I found was that there was a nice neat curve downwards from the higher classes to the lower ones, with about 0.4 seconds (two points or two lengths) separating one class from the next. I hadn't expected that, but was delighted that it happened.

We now take this a stage further and turn these figures into par figures. We start with a base figure of 100 (there is no significance in this number, it's just a neat starting point) and the Class 4 average was 5.2 points better (rounded down to five), so if we add that to 100 we get a par figure for Class 4 of 105. The Class 5 average was 3.125 points better (rounded down to three) so if we add that to 100 we get a par figure for Class 5 of 103. In my original calculation all those years ago, the Class 2 par did indeed come out at around 109, the Class 3 par at about 107 and the Class 6 par around 101. I merely extended these findings for the other classes that were not represented at the time, so that I ended up with the following pars –

Race Type	Par	Race Type	Par
Class 1 (Group 1)	117	Class 3	107
Class 1 (Group 2)	115	Class 4	105
Class 1 (Group 3)	113	Class 5	103
Class 1 (Listed)	111	Class 6	101
Class 2	109	Class 7	99

Through some more research I also found the following adjustments to be necessary for younger horses, depending on the time of year –

	2yo	3yo
January/February		-6
March/April	-11	-5
May/June	-10	-4
July/August	-9	-3
September/October	-8	-2
November/December	-7	-1
Races confined to fillies/mares		–3

These allowances are cumulative, so for example the par figure for a Class 4 event restricted to three-year-old fillies in July would be 99 (105 – 3 - 3).

Now that the par figures had been established, a real All-Weather meeting could be tackled and this is how it works. The following is a hypothetical six-race card with the distance of each contest shown.

The first thing to do is to take the difference between the winner's time in each race and the median time for each race distance, such as –

	Distance	Compared To Median Time
Race 1	7f	fast by 1.6
Race 2	6f	fast by 0.9
Race 3	1m	fast by 1.0
Race 4	1m2f	slow by 0.4
Race 5	1m4f	slow by 0.2
Race 6	5f	fast by 0.6

As I alluded to earlier, on the basis that the margins between the winning times and the median times are greater the longer the trip, I adjust all the margins shown above as though all races were run over a mile. Hence the margin of 1.6 shown for Race 1 is multiplied by 8/7, which gives a value of 1.8 (rounded). In other words, at the same velocity the margin between the winning time and the median time would have been 1.8 seconds had the race been over a mile and not seven furlongs.

Doing the same for the other races would give –

	Distance	Comparison Adjusted To 1m
Race 1	7f	fast by 1.8
Race 2	6f	fast by 1.2
Race 3	1m	fast by 1.0
Race 4	1m2f	slow by 0.3
Race 5	1m4f	slow by 0.1
Race 6	5f	fast by 1.0

At this stage, a raw speed rating is allocated to each winner based purely on the differences shown above. As one length equals 1/5 of a second, a difference of 1.8 seconds equals nine lengths or nine points on my scale. This number is then added to the base figure of 100 and that will result in a raw figure of 109 for Race 1. Fractions are always rounded down, so doing the same for the whole card would end up with –

	Distance	Raw Speed Figure
Race 1	7f	109
Race 2	6f	106
Race 3	1m	105
Race 4	1m2f	98.5 (rounded down to 98)
Race 5	1m4f	99.5 (rounded down to 99)
Race 6	5f	105

These raw ratings are then compared to the par for each class of race like this –

	Class	Par	Raw	Diff	Diff (Integer)
Race 1	5	103	109	fast by 6	-6
Race 2	4	105	106	fast by 1	-1
Race 3	1 *Listed*	111	105	slow by 6	+6
Race 4	2	109	98	slow by 11	+11
Race 5	3	107	99	slow by 8	+8
Race 6	6	101	105	fast by 4	-4

The next stage is where you have to use your judgement. The automated computer program I use will omit the races that were run fastest compared to par (in this case Race 1) and the slowest (Race 4) and then average the rest.

In this case that would be (–1 plus +6 plus +8 plus –4) divided by 4, which becomes 9 divided by 4. That would result in a track variant of 'slow by 2.25' (or +2.25) rounded to +2. That track variant figure would be added to the raw speed figures to give the final winners' speed figures, on the basis that the conditions had slowed the horses down by two lengths for every mile they ran.

Note: If the differences between the par and raw speed figures congregated around one figure, even if that was different to the average produced by the automatic system described earlier, I would be likely to settle on that. This part of the speed figure compiling process is the most subjective and requires you to make a judgement using all the skills and knowledge available to you.

However, if I am happy that in this case the track variant of +2 is accurate, then that figure is added to the raw speed figures for each winner to give the final speed figures. Had the track variant been negative (meaning faster conditions) then that figure would have been subtracted from the raw figures to provide the final ratings. Therefore the winners' speed figures for this meeting would be –

	Class	Raw	Variant	Final Speed Figure
Race 1	5	109	+2	111
Race 2	4	106	+2	108
Race 3	1 *Listed*	105	+2	107
Race 4	2	98	+2	100
Race 5	3	99	+2	101
Race 6	6	105	+2	107

The way that the speed figures for the beaten horses are calculated uses similar methods, in that the number of lengths the horse in question finished behind the winner is adjusted to take into account the distance of the race (it is again adjusted to one mile). For the beaten horses in Race 3, which was over exactly a mile, the calculation is simple. If the second horse had finished six lengths behind the winner and the winner had been given a rating of 107, then the runner-up will be awarded 101 (107-6). However, for a horse that had also finished six lengths behind the winner in Race 5, he would earn a rating only four points lower than the winner's rating of 101, because the race was over a mile and a half (or 12 furlongs). In this case the calculation would be -

$$101 - (8 / 12) \times 6$$

or $$101 - 4 = 97$$

The only thing that now has to be decided is whether you make adjustments to the speed figure based on the weight the horse carried, and then eventually adjust that figure to account for the weight the horse is set to carry when it races again. Personally I don't do this, as I want to leave the speed figure as just that – a speed figure which has had class and going taken into account, but not weight. This has to be a personal decision and I don't make adjustments purely because I have never quite been able to work out to my satisfaction how much of an effect that an extra pound added on or taken off has on the speed of a horse. My own research over the years has only produced ambiguous results, so I prefer to use added or reduced weight as what I call a "hunch" factor, rather like going requirements, fitness, stamina, jockey, trainer, etc, rather than distorting the speed figures so that you can lose sight of who the genuinely fastest horses are.

This entire process may all seem a bit confusing if you aren't familiar with speed figures and how they are calculated, but experience has shown that the time spent going through all of this has been repaid many times over. What you end up with is a way of opening a race up and providing you with a point at which to start tackling the contest involved. Whether you use published speed ratings or calculate your own, the eye is naturally drawn straight away to those horses with the higher ratings and one thing you can be sure of is that they have run at least one fast time (allowing for the conditions) at some point in their career and the proof is there in front of you.

My own preference is to look for horses that consistently earn high speed figures under a particular set of circumstances, i.e. over the same trip or at the same track, or when they lead early, and they don't necessarily have to have won the races in question. In fact I don't automatically assume that a horse is suited by certain conditions simply because it has won a race under them as it may have earned a low speed figure in doing so. Any horse can win a race if it outclasses the opposition and a modest speed figure can be a sign that this has happened in a slowly run race.

The other way I use speed figures, and this will be familiar to readers of King Of The Sand in the *Racing & Football Outlook*, is to make a note of horses whose speed rating way exceeds the par for the class of race they were running in. Again they don't necessarily have to have won, but if they were to be awarded a speed rating of 107 in a Class 6 contest (par 101), wherever they may have finished, they would automatically be placed in a special listing on my computer. This forms the basis of the *Hotlist* in the King Of The Sand column.

Note: I can present my own speed ratings in two different formats. The full version containing many more factors than just the speed figures themselves will be shown later on, but the summary version appears like this.

2-20 CALL PONTIN'S NOW 0844 815 3648 CLAIMING STAKES (6) 4YO+
7 furlongs — Par = 101 — 4 runners

AW FORM NAME	BEST	LAST SIX
15119-13 LOUISIADE	108S07	95L08 108S07 99W08 88S08 95S07 98W08
9006-048 CAPITAL LASS	107S07	53S07 69S07 77S07 85S07 90S07 75W07
31977-31 ANOTHER GENEPI	105W07	102W07 89W07 94W06 94W07 99W07 102S07
2987-921 GOVERNMENT	97S07	89W06 91S07 89S08 93S07 97S07 95S07

This version shows a limited number of factors, but they can be quite valuable even in this format. Certainly there is more information here than will be available to 99% of punters and bookmakers, the very people you will be betting against. Most of the details shown here will be fairly self-explanatory. The form figures shown under "AW Form" are for the horse's most recent starts on sand only, up to a maximum of eight outings with the most recent on the right. After the horse's name and under "Best" is the horse's best speed figure on sand in its entire career, together with the track and the race distance.

Under "Last Six" are the horse's last six speed figures on sand, the most recent on the right, together with the track and the race distance. The best and most recent speed figures are shown in this format.

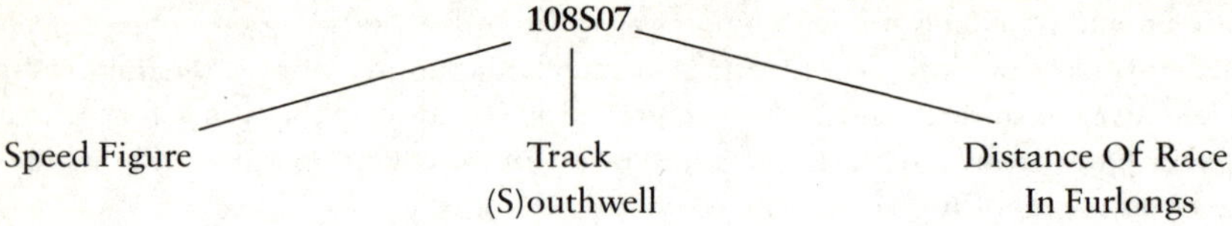

Quite apart from giving an insight into how suited the horse is likely to be by the conditions (provided they have encountered something like them within their last six outings) this type of presentation will also tell you whether the horse is likely to be fast enough for the type of race they are running in. The answer is yes for three of the four runners in this race as they had run faster than the class par at some point in their lives, though only two of them had done so within their last six outings, namely Louisiade and Another Genepi. Incidentally both had done so over the same course and distance as for this race, but as things turned out this was to be Louisiade's day with the 5-2 shot just getting up to defeat the 4-9 favourite Another Genepi.

In summary, speed ratings are a very valuable tool on the All-Weather, but even so they are merely part of a much bigger puzzle. There are many more pieces to be added yet, as we shall see.

ANALYSING FORM

I THINK everyone knows what we mean by the word "form". It's that intangible thing that goes out of the window every time a horse we believe has no chance of winning goes and wins. A more sensible definition is that it means a sort of a signpost or a record of a horse's recent placings in numerical form, though it can mean much more than that. Interestingly, amongst the alternative phrases for the word "form" according to the Oxford Thesaurus is "to fabricate", but that's another story.

For the purposes of this book, the term "form" is taken to mean the parts of a horse's past performances that are most relevant to what is likely to happen in the future. These relevant parts need to be extracted if a proper conclusion is to be made, and most newspapers and other publications that contain racecards will at least sift the form in the most basic of ways by bolding form figures for past races run on sand.

For example, this horse was due to take part in a one-mile-four-furlong seller on the Lingfield Polytrack on June 28th 2008 and this is how he would have appeared in most publications.

2-11143	Bridgewater Boys	9-9-13	G L Moore	Jemma Marshall (7)

Even in this most basic of formats, it's easy to see that this horse is a sand specialist. Those form figures in bold type stick out like a sore thumb and anyone with little or no knowledge of the game could pick him out, provided they were able to understand the difference between bold and plain form figures. Naturally Bridgewater Boys won and was obviously expected to do so as he went off the 10-11 favourite, but that's not the point. The point is that you need to be able to sort out form shown on sand from form shown on grass, for it is previous form on sand that is the biggest of factors when it comes to assessing an All-Weather race.

There are many ways to do it, from the basic sifting of bold form figures shown above to the use of electronic formbooks such as *Raceform Interactive*, which makes the whole process much easier. The way I have done it is to put together my own All-Weather database (examples from which will appear at various stages later on) and as impressive as *Raceform Interactive* is, I still would have done so even if the software had been available before I started to create my own database. The reason is that I can manipulate the data any way I want to and attach my own little bells and whistles to it. The only question is to what extent should turf form be taken into account and where should the line be drawn?

In the old days of sand racing it was quite straightforward. You could basically ignore any form shown on turf when analysing an All-Weather contest, unless a recent run on grass guaranteed a horse's fitness. The old Equitrack surface at Lingfield and Fibresand surfaces at Southwell and Wolverhampton were so specialised that all you needed to do was focus on those horses that had already run well on the relevant surface. Time after time people fell into the old trap of backing horses whose best form had been shown on turf, only to see the animal in question cough and splutter its way around the circuit, choking on the dust kicked up by the All-Weather specialists and lifting a hoof up to shield its eyes from the kickback – metaphorically speaking.

One of the earliest and best examples of this came in the 1997 Wulfrun Stakes at Wolverhampton in which Fulke Johnson Houghton's classy performer Centre Stalls was backed down to 6-5 favourite. Officially rated 116 at the time, earlier in the year he had finished third behind Bosra Sham in the Brigadier Gerard Stakes at Sandown and second behind Allied Forces in the Queen Anne Stakes at Royal Ascot. He would surely be too good for his six rivals at Dunstall Park then, wouldn't he? Well he probably would have had this race been run on grass, but it wasn't, it was on Fibresand and he hated it, eventually finishing a long last behind the winner Farmost and a certain horse called Running Stag. The official explanation as shown in the official formbook blandly said it all –

OFFICIAL EXPLANATION: Centre Stalls (IRE): was unable to act on the surface.

This was probably one of the first, and certainly up to that point, the highest-profile example of how different turf and sand were. Admittedly the jump from racing on turf to racing on the Wolverhampton Fibresand was a major one; too much for Centre Stalls as it turned out, but he was certainly not alone. Horses like him were meeting the same brick wall day in day out. The only reason they didn't generate the same publicity was because the races involved were at a much lower level. No doubt some people believed that the reason so many horses with apparently impressive credentials from the turf were flopping on sand was because they were modest horses by definition, and modest horses are unreliable. There was never any suggestion they were just not good enough.

So have things changed that much in the past 11 years? Well yes they have, at least in some areas. The arrival of Polytrack has certainly made it easier for turf horses to make the transition from grass to sand, but the reasons for that are debatable. One theory is that Polytrack makes it easier for jockeys "to ride a race", or more accurately it means they can ride a turf race on sand, i.e. drop out and come late which was a very hard thing to do on the older sand surfaces. I also believe that the reason so many more horses are clicking on Polytrack at the first time of asking is that in general better horses are appearing on sand these days, hence the better races now being run on it (on Polytrack at least). Back in 1997 the Wulfrun Stakes was the only Listed race run on an artificial surface all year in Britain, but now there are countless numbers of them and there are even three Group 3 races, so it's easy to see how the standard has risen.

Despite all of that, I still feel as though I'm walking into the racecourse with no trousers on (a horrid thought) if I haven't got my own sheets of All-Weather form about my person and it's very rare indeed for me to back a horse on Polytrack that has no previous sand form to its name. To me doing so would be pure guesswork.

That's the way I see Polytrack racing, but the situation for Fibresand is the same as it's always been. Not only would I be very reluctant to back a horse on Fibresand that had no experience on that surface, I would also be dubious about backing horses with proven form on Polytrack when they first run at Southwell. "Fibresand Horses For Fibresand Races" is my motto, but there are exceptions of course. Enough racing still takes place on Fibresand to be able to sit out any races where there are any doubts as to what might happen. If you know what you're doing, however, there is plenty of money to be made on Fibresand when the right sort of opportunity arises. Here is a perfect example.

The Southwell fixture on Wednesday March 26th 2008 included a one-mile-four-furlong Class 4 handicap, for which a field of nine was due to go to post. Amongst the runners was what appeared to be a real handicap "blip" in the shape of Clive Brittain's Sahrati, the mount of Robert Winston. The reason he looked such a steal (and the reason why he went off as short as 13-8 for what was quite a decent little handicap) was that he was racing off a 17lb lower mark than on turf. He was returning from six months off, but that wasn't too much of a problem considering how much he appeared to have in hand. What was a problem, however, was that he did have previous "form" on Fibresand, and it was bad.

In fact he had finished a tailed-off last of four when an even-money favourite for a nursery at Southwell right at the end of his two-year-old season. He had performed a little better in a couple of races on Polytrack soon afterwards and went on to win twice on turf in 2007, but that couldn't disguise the fact that he still had a major question mark against his ability to handle the surface. The evidence of that solitary previous effort on Fibresand should have been enough to sound the warning bells for anyone thinking of backing him. Of course anyone who had bothered to look back far enough in his form, which is obviously easier if you have a computerised formbook or you maintain your own All-Weather database like I do, would have seen at a glance that there was a glorious opportunity awaiting here.

The opportunity to lay horses these days made this an especially attractive prospect and that is what I did, but for some reason – and to this day I still don't know why – I didn't lay him for a place. Surely given the doubts about him I should have done so (and to decent money too) as to my mind he was either going to put his previous bad effort on Fibresand behind him and show his true ability this time, or he was going to flop again. In other words I couldn't really imagine him finishing second or third – he was either going to win or he was going to bomb out completely.

I watched the race with great interest and felt my opinion to have been well and truly vindicated when he came off the bridle fully half a mile from home, found absolutely nothing, and eventually trailed in seventh of the nine runners, beaten over 24 lengths behind the winner Tartan Tie. I hadn't backed the winner, I backed the runner-up Bentley Brook, but although I had "won" by laying Sahrati for a win, not laying him for a place meant that I didn't make anything like as much money as I should have. The postscript to the race once again appeared beneath the result as shown in the official formbook –

OFFICIAL EXPLANATION: Sahrati – jockey said the colt never travelled.

Of course there are plenty of horses that are the complete opposite of Centre Stalls and Sahrati in that they only show their best form on sand and are nothing like as good on turf. Usually this is reflected by their official ratings and a good example of this is a horse called Gentleman's Deal.

A huge individual for a Flat horse as anyone who has stood close to him will know, he did win on turf and on Polytrack in three starts for Ed Dunlop as a three-year-old, but his Chepstow maiden win came at the main expense of a horse officially rated 42, hardly stellar form. However, when he returned from a 15-month absence at the start of 2006 in the care of Mick Easterby, things really started to take off. He won twice on sand, at Southwell and Wolverhampton, and despite an 8lb rise in his handicap mark he went off at short as 10-1 for the Lincoln, that year run at Redcar.

This was the first real sign that he was a true sand specialist, as he trailed in 21st of the 30 runners, beaten over 27 lengths behind the winner Blythe Knight. Five more outings on grass followed that year and he failed to make the first three in any of them, but it was a different story when he returned to sand after the turf season ended, as he soon rattled up a four-timer culminating in a famous victory in the 2007 Winter Derby at Lingfield. His official mark shot up from 91 to 108 in that period, so when it was announced that he would turn out again quickly for another crack at the Lincoln (that year run at Newcastle) off a turf mark of 93, he quickly became a popular choice for many and was eventually sent off the 7-2 favourite.

It seemed that the lessons of the previous year hadn't been learned though. The ground was very testing once again, just as it had been at Redcar, and he hadn't run his race there so why should he do any better this time? It appears that a lot of people became obsessed by the fact that he was racing off a 15lb lower mark than on sand and his form figures coming into the race would have looked very impressive, even if they were in bold!

Gentleman's Deal maintains his unbeaten record on sand with victory in the 2007 Winter Derby

The outcome was sadly predictable. Gentleman's Deal trailed in 15th of the 20 runners, 24 lengths behind the winner Very Wise. The following official explanation wouldn't have been the greatest surprise –

> **OFFICIAL EXPLANATION: Gentleman's Deal (Ire) – jockey said the horse was unsuited by the ground.**

Up until the end of 2007, Gentleman's Deal's form figures on turf were 4109454700, whilst his profile on sand was 111111121. Is any more evidence needed?

There are plenty of similar horses to Gentleman's Deal at a rather more modest level of course, but the value with any horse like him, irrespective of their class, is when they return to their favoured surface having been struggling on a less-favoured one. This also applies to horses moving between Polytrack and Fibresand as well and there are examples of this later on in the book.

Here is an example of the sort of horse worth looking out for. Andrew Balding's gelding Border Music was due to line up for a seven-furlong Listed race at Wolverhampton on March 10th 2007 and his form figures coming into the race as shown in most newspapers, and in the racecard, were 167227, creditable enough but nothing to write home about. He also looked to have it to do against one of his main rivals in the shape of Jack Sullivan, as he was officially rated 7lb inferior to him. However, strip away the turf runs and his last six form figures would have been 411111, which puts a rather different complexion on things. In fact, looking at his profile as it appeared in my own personal All-Weather racecard showed just how much of a versatile sand-specialist he had become.

02 NOV 2003	LING	8	SW	4MDN	1	10	9-0	4	H	---	97
19 MAR 2004	LING	7	SD	4CLF	7	8	8-3	6	H	---	99
27 APR 2005	LING	8	FT	4HCP	4	12	9-11	10	P	---	106
08 AUG 2005	WOLV	7.1	FT	5HCP	1	12	9-9	1	L	---	108
27 OCT 2005	LING	7	SD	2HCP	1	13	8-9	14	P	---	108
30 OCT 2005	LING	6	SD	2HCP	1	12	8-13	1	H	---	104
11 MAR 2006	WOLV	7.1	FT	2CND	1	5	9-0	5	L	---	107
01 APR 2006	KEMP	6	VS	3CND	1	6	9-10	5	P	---	109

Note: the finishing position is the sixth column.

Despite an impressive-looking profile like this, he was allowed to go off a 15-2 shot, but showed what he thought of that by beating Jack Sullivan convincingly and shaving half a second off the course record. Incidentally, Border Music did manage to win his first race on turf three months later, but it had taken him 25 races on grass to do so.

Previous form and the surface it has been shown on is one thing, but finding the right type of race to bet in is another. Gentleman's Deal and Border Music are examples of the classier type of animal, the type who contest Pattern events or at the very least decent conditions events on sand. Personally I find such races difficult and confusing when it comes to betting on them, mainly because these contests often attract classy performers off the turf and it's hard to know what to expect from them, though opportunities do arise thanks to the likes of Gentleman's Deal and Border Music.

At the other end of the scale I am no great fan of sellers or claimers as the horses that take part of them are either considered badly handicapped by their connections, are generally on the decline, or are just plain moderate. You often find horses rated in the 80s taking on others rated in the 40s or 50s. The classier horse often wins, but usually at a ridiculously short price. Everyone knows they will probably win anyway, so there is no real edge for the shrewd All-Weather form expert to exploit.

The old banded races, now re-branded as Class 7 classified events, can produce a few opportunities, as unlike in sellers and claimers the horses taking part in them are all around the same (very low) official mark and there may be one or two lurking in there that will prove better than the grade. I don't really like maiden events or just about any race confined to younger horses though. By their very nature they are contested by newcomers or unexposed sorts with little previous form of any sort, let alone on sand.

That leaves handicaps for older horses and in my opinion they are the races to concentrate on. In most cases they are competitive, the betting market is relatively open, and the horses taking part in them have often established some sort of preferences within their profiles. The good news is that there are enough of them to be able to leave the other aforementioned types of races to the greedy and the desperate.

Here are a couple of horses whose liking for a particular set of conditions are well proven, one of them exposed, the other less so.

The first is a horse called Garlogs.

21 NOV 2005	SOUT	7	VS	6SEL	6	13	9-3	6	P	---	73
06 DEC 2005	SOUT	5	FT	4HCP	3	11	7-5	6	L	---	98
15 DEC 2005	SOUT	5	VF	5HCP	2	12	7-5	9	P	---	94
16 DEC 2005	WOLV	6	FT	4HCP	6	12	7-5	9	P	---	95
12 JAN 2006	SOUT	5	SD	6HCP	1	8	9-1	6	P	---	95
07 FEB 2006	SOUT	5	FT	6HCP	1	9	9-4	1	L	---	95
09 FEB 2006	SOUT	5	VF	5HCP	2	8	8-10	3	L	---	92
10 FEB 2006	WOLV	5.1	FT	6HCP	2	6	9-10	4	L	---	94
20 MAR 2006	SOUT	5	SD	5HCP	2	7	9-2	1	L	---	93
11 NOV 2006	WOLV	5.1	FT	5HCP	12	12	9-1	11	P	---	68
24 NOV 2006	WOLV	5.1	FT	5HCP	12	12	8-11	11	P	---	90
27 NOV 2006	WOLV	5.1	SD	6HCP	8	12	9-1	1	L	---	96
02 JAN 2007	SOUT	5	VF	6HCP	4	13	8-12	11	P	---	97
01 FEB 2007	SOUT	5	SD	6HCP	3	10	9-0	12	P	---	97
15 FEB 2007	SOUT	5	VS	5HCP	3	11	8-2	4	P	---	102
22 FEB 2007	SOUT	5	SW	6HCP	2	9	8-9	9	P	---	95
06 MAR 2007	SOUT	5	SD	5HCP	6	11	8-4	8	L	---	95
13 MAR 2007	SOUT	5	SW	6HCP	3	13	9-2	2	L	---	95
03 MAY 2007	WOLV	6	FT	6HCP	6	13	8-12	9	P	---	98
14 MAY 2007	WOLV	5.1	FT	6CLM	6	12	9-8	13	P	---	94
12 JAN 2008	WOLV	5.1	SW	6HCP	1	8	8-6	10	L	-5	103
25 JAN 2008	WOLV	5.1	SD	6HCP	4	6	8-5	4	L	+3	98
17 FEB 2008	SOUT	5	VS	6HCP	1	12	8-9	8	L	---	101
04 MAR 2008	SOUT	5	SD	5HCP	1	8	8-12	3	L	-6	103
20 MAY 2008	SOUT	5	SD	4HCP	1	8	8-4	3	P	-28	105

Garlog's finishing positions for his races on sand (sixth column from left) will show that over the straight five furlongs at Southwell, his form figures are 321122433263111 and it is only when you see a horse's previous form on sand shown like this that everything becomes much clearer. Incidentally, one of the reasons he is so suited to these conditions is explained in the third column from the right, which is a 'pace' symbol. This is a single character which describes the position the horse concerned took early in the race. The letter 'L' means that he led early, whilst 'P' means that he was prominent early. The letter 'H' means 'held up' or basically anything not covered by the other two symbols (the subject of early pace will be covered much more fully later on).

The front-runner statistics shown in the chapter on Southwell show just how important early speed is over the straight five furlongs there, so it's hardly a surprise that Garlogs has such a good record under those conditions.

Garlogs is an experienced sprinter who regularly contests just the sort of older-horse handicaps that I have been talking about, and there are plenty more like him.

By comparison Gala Evening is much less experienced on sand, but he has an interesting profile nonetheless.

07 MAY 2007	KEMP	16	SW	5HCP	1	13	9-10	7	H	---	100
12 APR 2008	KEMP	16	SD	4HCP	1	8	9-11	2	H	+7	110
22 APR 2008	KEMP	16	SD	4HCP	1	5	9-11	5	P	+8	106

All-Weather racing at Kempton is still in its infancy compared to the likes of Lingfield, Southwell and Wolverhampton, yet here is a horse that has already taken to the venue like a duck to water. The real surprise here is that for the second of those three victories (on 12th April 2008) he was allowed to go off at 20-1. Why? Because during the 11 months which had elapsed since his first Kempton victory he had run four times – once on turf and three times over hurdles – but was beaten out of sight each time. A little less obsession with form shown under very different disciplines could have put a few more people on to a big-priced winner, but it still goes to show the rewards that can be found if you delve a little deeper.

He was only a 2-1 shot when he won his third race over two miles at Kempton ten days later, but he didn't even start favourite for that!

PACE BIASES

HOW OFTEN do you hear racing professionals say that a track was suiting front-runners, or that a pace bias was in operation? I have used such phraseology many times myself, but believe it to be of huge importance.

We can all recognise a pace bias when we see it unfolding, such as when those horses that lead early seem to be making all race after race, but that's only half the battle. The other half is being able to find out whether a horse was inconvenienced by a pace bias in its past performances, or even better, find those who have put up a better effort than it might have appeared in the face of a pace bias, rather like identifying horses that have performed well against a draw bias.

Quite often pace biases are perceived without any statistical data to back them up. For example, you may see something like: "Horse 'A' did well to come from off the pace and finish so close on a track that was suiting front-runners." The problem is that amongst most racing publications, formbooks etc, it's not easy to find out how a track was riding with regards to a pace bias alongside all the other information shown within the past performances, at least not in an instantly recognisable format. There needs to be a way of recording this data for each meeting and I have been forced to do the work required for myself, but it has been rewarding and it has also been profitable.

The result of all of this work is that I have been allocating what I call a 'pace rating' for every All-Weather meeting since 1st January 2007. This is basically how it works.

I look at the comments-in-running for the first three horses home in all races at a particular meeting. Depending on the early position they took in the contest I allocate –

a) The highest figure to any winner that had led early.
b) A smaller figure to an early leader who then finished second or third, or any horse that was prominent early and then managed to go on to win.
c) A lower figure to any horse that was prominent early and then finished second or third.

Nothing is added if the horse doesn't meet any of the above criteria. The size of each set of figures, and therefore the scale used, is up to you, but the way I do it is to allocate 200 points to each qualifier from Group A, 100 points to those from Group B, and 50 points to those in Group C. Finally, the figures allocated are then all added together and divided by the total number of horses to make the first three in all races at the meeting.

I always find that a scale of 0 to 100 is very easy to interpret, so I have designed the above system in such a way that if all the early leaders won at a six-race meeting, and the 12 placed horses were all prominent early, the pace rating would come out at 100 which in theory is the maximum possible.

Note: You get this result because all six winners would earn 200 points each and all 12 placed horses 50 points each, so the total would be 1800. The number of horses who made the first three in all six races is 18, and 1800 divided by 18 is 100.

Admittedly if two horses led very early in one race, and one of them won whilst the other made the frame, the pace rating could in theory be higher than 100, but that is about as likely as the theoretical highest break of 155 (including a free ball) being achieved in snooker.

If all 18 horses that finished in the first three on a six-race card were held up early, the pace figure would be zero.

I make sure that the pace rating is visible in the all-in-line form I use for All-Weather racing, so then I can see at a glance if a performance was influenced one way or another by a pace bias. This is how I interpret the resulting pace ratings.

Pace Rating	
0-20	Hold-up horses held an overwhelming advantage and those that raced up with the pace might as well have stayed at home.
21-40	Significant advantage to hold-up horses.
41-60	No great advantage either way.
61-80	Significant advantage to prominent racers.
81 or above	Prominent horses held an overwhelming advantage and hold-up horses might as well have stayed at home.

This sort of system can be used for any sort of racing, not just on sand. Here is an example of the system in action.

Whilst assessing the form for the five-furlong handicap at Wolverhampton on 11th June 2007, I had come to the conclusion that Monte Major had an outstanding chance on most of his recent form, but he had appeared to run moderately at Kempton just five days earlier. However, according to my pace ratings any horse that tried to come from behind, as Monte Major did, at the Kempton meeting on June 6th faced a near-impossible task.

In fact the pace rating I awarded for that meeting was the highest that I had allocated to any All-Weather meeting up until then. The actual figure was 81, and a glance at the above scale shows just exactly what Monte Major was up against. Having found a genuine reason to forgive Monte Major his Kempton effort, I followed the courage of my convictions and was rewarded with an easy 5-1 winner.

Interestingly, the opening race at the June 6th meeting was an ordinary-looking two-year-old median auction maiden event over five furlongs, in which the favourite Concertmaster made all of the running and nothing was able to get into the race from off the pace. Amongst those that tried was a filly called Little Knickers who was making her debut and she stayed on to finish sixth of the 11 runners despite being hampered, thereby showing a glimmer of ability. If you had made a note of how the Kempton Polytrack was riding that day and that any horse who made any sort of headway late on had run better than their finishing position would suggest, you might just have been on her when she went in at Bath next time at odds of 33-1.

At the other end of the scale is the meeting at Lingfield on October 29th 2007. Hold-up horses are generally at an advantage at Lingfield, especially over the longer distances, but at this fixture the bias was even greater than normal. All six winners were held up and only one horse that led early managed to finish in the first three. The pace rating for the meeting was just 19, which is the lowest pace rating figure I have allocated any fixture since I have been calculating them. It was impossible for prominently ridden horses to make any impact at all and by definition those that tried had to be forgiven. Amongst those that were ridden up with the pace at the meeting were three horses who all went on to win their next starts, namely Ramatni (3-1), Cinnamon Hill (5-1) and Satin Braid (10-1).

Obviously both the Kempton and Lingfield examples are extremes and the pace rating I award to the vast majority of meetings will fall into the 41 to 60 category, but that doesn't mean that I'm not always on the lookout for a bias, because if you can spot a valid one early enough you can make it pay. You can't draw any hard and fast conclusions about a meeting after just one race, but if after a couple of contests you have noticed that the principals in both (not just the winner) had either all been prominent or all held up, then you should be looking even harder to see what happens in race three, because it may be just a coincidence. If you commit yourself prematurely you may become unstuck.

If, however, the third race confirms your findings then that is the time to get stuck in, because you will still have at least three races left on the card and that should be enough time for anyone to make a bit of cash. Once the meeting is over, if you do what I do and make a note of the bias in some form or other (like a pace rating) you will also have an idea on how to treat the form at this fixture when the horses that took part turn out again.

A while ago I tried to work out whether it was possible to anticipate whether a pace bias might exist based on the weather and the apparent state of the track. There is a suggestion that race times speed up on All-Weather courses after rain has fallen and this makes sense if you imagine a beach after the tide has gone out. The sand there is wet and hard and if you run on it you will be able to move faster than on the soft, dry, powdery stuff further away from the water, where your efforts just result in more kickback.

If a similar thing happens on an All-Weather track, would that then result in a pace bias? The following information, taken from my database for the first three months of 2008, might shine some light on that. What this shows is, from left to right, the date of the meeting; the name of the track; the track's variant (a minus figure is faster than standard, a plus figure slower than standard); a description of the conditions based on the track variant, and finally the pace rating for the meeting.

LINGFIELD				
03 JAN 2008	LING	+1	Standard	38
05 JAN 2008	LING	-3	Standard	45
09 JAN 2008	LING	0	Standard	55
12 JAN 2008	LING	-3	Standard	39
16 JAN 2008	LING	0	Standard	60
19 JAN 2008	LING	+1	Standard	60
23 JAN 2008	LING	+6	Slow	33
26 JAN 2008	LING	0	Standard	50
30 JAN 2008	LING	+4	Slow	64
02 FEB 2008	LING	-2	Standard	76
04 FEB 2008	LING	-3	Standard	67
06 FEB 2008	LING	0	Standard	48
09 FEB 2008	LING	-1	Standard	71
13 FEB 2008	LING	0	Standard	31
15 FEB 2008	LING	+1	Standard	55
16 FEB 2008	LING	0	Standard	52
18 FEB 2008	LING	0	Standard	48
20 FEB 2008	LING	0	Standard	48
23 FEB 2008	LING	-1	Standard	55
26 FEB 2008	LING	+1	Standard	55
28 FEB 2008	LING	0	Standard	57
03 MAR 2008	LING	0	Standard	53

05 MAR 2008	LING	+1	Standard	64
14 MAR 2008	LING	+4	Slow	45
15 MAR 2008	LING	+3	Standard	44
19 MAR 2008	LING	+1	Standard	36
27 MAR 2008	LING	0	Standard	44
28 MAR 2008	LING	0	Standard	52
31 MAR 2008	LING	0	Standard	71

KEMPTON

02 JAN 2008	KEMP	-1	Standard	43
05 JAN 2008	KEMP	-1	Standard	55
09 JAN 2008	KEMP	0	Standard	64
11 JAN 2008	KEMP	-4	Fast	57
13 JAN 2008	KEMP	+1	Standard	43
15 JAN 2008	KEMP	+9	Slow	38
16 JAN 2008	KEMP	-2	Standard	40
18 JAN 2008	KEMP	0	Standard	57
21 JAN 2008	KEMP	+6	Slow	69
23 JAN 2008	KEMP	0	Standard	57
25 JAN 2008	KEMP	-2	Standard	62
27 JAN 2008	KEMP	-2	Standard	38
28 JAN 2008	KEMP	-1	Standard	29
01 FEB 2008	KEMP	+5	Slow	48
02 FEB 2008	KEMP	+3	Standard	38
03 FEB 2008	KEMP	0	Standard	48
06 FEB 2008	KEMP	+1	Standard	38
10 FEB 2008	KEMP	0	Standard	52
13 FEB 2008	KEMP	+4	Slow	41
16 FEB 2008	KEMP	+3	Standard	45
20 FEB 2008	KEMP	+1	Standard	55
24 FEB 2008	KEMP	+4	Slow	52
27 FEB 2008	KEMP	-1	Standard	64
28 FEB 2008	KEMP	-2	Standard	69
29 FEB 2008	KEMP	+2	Standard	63
14 MAR 2008	KEMP	-2	Standard	36
15 MAR 2008	KEMP	-4	Fast	50
18 MAR 2008	KEMP	0	Standard	64
19 MAR 2008	KEMP	-1	Standard	60
22 MAR 2008	KEMP	-2	Standard	71
28 MAR 2008	KEMP	+2	Standard	45
29 MAR 2008	KEMP	-5	Fast	43

SOUTHWELL

01 JAN 2008	SOUT	+1	Standard	64
04 JAN 2008	SOUT	+6	Slow	63
06 JAN 2008	SOUT	0	Standard	48
07 JAN 2008	SOUT	+2	Standard	57

08 JAN 2008	SOUT	+3	Standard	55
10 JAN 2008	SOUT	+1	Standard	83
14 JAN 2008	SOUT	-2	Standard	64
15 JAN 2008	SOUT	0	Standard	56
17 JAN 2008	SOUT	+4	Slow	43
22 JAN 2008	SOUT	-4	Fast	67
24 JAN 2008	SOUT	-4	Fast	50
29 JAN 2008	SOUT	-4	Fast	60
31 JAN 2008	SOUT	+2	Standard	64
04 FEB 2008	SOUT	-3	Standard	58
05 FEB 2008	SOUT	+2	Standard	60
07 FEB 2008	SOUT	-2	Standard	44
08 FEB 2008	SOUT	-5	Fast	33
12 FEB 2008	SOUT	+4	Slow	81
14 FEB 2008	SOUT	+11	Very Slow	72
17 FEB 2008	SOUT	+16	Very Slow	61
19 FEB 2008	SOUT	+18	Very Slow	56
19 FEB 2008	SOUT	+18	Very Slow	50
21 FEB 2008	SOUT	+10	Very Slow	72
22 FEB 2008	SOUT	+12	Very Slow	67
04 MAR 2008	SOUT	+12	Very Slow	53
11 MAR 2008	SOUT	+12	Very Slow	42
12 MAR 2008	SOUT	+12	Very Slow	48
13 MAR 2008	SOUT	+6	Slow	50
20 MAR 2008	SOUT	-2	Standard	67
26 MAR 2008	SOUT	-1	Standard	54
31 MAR 2008	SOUT	+12	Very Slow	50

WOLVERHAMPTON

02 JAN 2008	WOLV	+9	Slow	52
03 JAN 2008	WOLV	+6	Slow	47
04 JAN 2008	WOLV	+7	Slow	58
09 JAN 2008	WOLV	+12	Very Slow	42
10 JAN 2008	WOLV	+10	Very Slow	61
11 JAN 2008	WOLV	+11	Very Slow	55
12 JAN 2008	WOLV	+7	Slow	56
17 JAN 2008	WOLV	+7	Slow	56
18 JAN 2008	WOLV	+6	Slow	35
19 JAN 2008	WOLV	+9	Slow	61
20 JAN 2008	WOLV	+9	Slow	57
21 JAN 2008	WOLV	+7	Slow	55
24 JAN 2008	WOLV	+2	Standard	33
25 JAN 2008	WOLV	+1	Standard	48
26 JAN 2008	WOLV	+4	Slow	58
28 JAN 2008	WOLV	+6	Slow	45
30 JAN 2008	WOLV	+10	Very Slow	61
31 JAN 2008	WOLV	+12	Very Slow	44

01 FEB 2008	WOLV	+14	Very Slow	40
07 FEB 2008	WOLV	+10	Very Slow	53
08 FEB 2008	WOLV	+9	Slow	42
09 FEB 2008	WOLV	+6	Slow	61
11 FEB 2008	WOLV	+2	Standard	60
14 FEB 2008	WOLV	+8	Slow	47
15 FEB 2008	WOLV	+7	Slow	58
21 FEB 2008	WOLV	+6	Slow	64
22 FEB 2008	WOLV	+9	Slow	42
23 FEB 2008	WOLV	+5	Slow	67
25 FEB 2008	WOLV	+6	Slow	67
27 FEB 2008	WOLV	+9	Slow	79
29 FEB 2008	WOLV	+11	Very Slow	55
01 MAR 2008	WOLV	+7	Slow	42
03 MAR 2008	WOLV	+4	Slow	48
05 MAR 2008	WOLV	+9	Slow	47
06 MAR 2008	WOLV	+9	Slow	47
07 MAR 2008	WOLV	+10	Very Slow	50
08 MAR 2008	WOLV	+5	Slow	53
12 MAR 2008	WOLV	+4	Slow	56
13 MAR 2008	WOLV	+4	Slow	58
17 MAR 2008	WOLV	+3	Standard	55
22 MAR 2008	WOLV	-1	Standard	56
26 MAR 2008	WOLV	0	Standard	47
27 MAR 2008	WOLV	0	Standard	53
28 MAR 2008	WOLV	0	Standard	61
29 MAR 2008	WOLV	+9	Slow	67

What I was hoping to find was that if the surface was significantly riding one way or the other, i.e. very fast or very slow, then that would result in a proper bias. Unfortunately, as hard as I looked I couldn't find a discernible link between the speed of the track and a pace bias. I had thought that perhaps when the weather became very cold or very wet, and the track became compacted as a result, then there would be bias towards closers as they would have a better chance to accelerate on a more solid surface, but there is no evidence of that. Perhaps if the track is deep harrowed in such adverse conditions, it makes the surface more holding and gluey and therefore more difficult to change speed on.

Either way, the conclusion is that it's best to wait until there is enough evidence of a bias being in operation before you try and capitalise on it. Once the meeting is over you should make a record of any bias for future reference, as it might influence how you interpret the form shown at the fixture.

TRACK BIASES

TRACK BIASES are rather more difficult to fathom than pace biases, at least when looking at a horse's past performances, and the reason is quite simple. You can immediately see what position a horse took in the early stages of a race by looking at the comment-in-running as shown in the form section of the *Racing Post* or in the official formbook, such as: "soon led", "tracked leaders" or "held up in rear", and then allocate a pace symbol ('L, 'P' or 'H') but you can't be so certain where these horses were in relation to the inside rail. For instance, was he/she tight against it or six horses deep? When they made headway, was it up the inside or on the wide outside?

I should say that the race-readers who work for the various publishers of racing results do a fine job and they will try and point out where the horse raced or made its effort when they can. They will say things like: " effort on wide outside over 1f out", or "tracked leaders on inside" and this is obviously of significance.

Part of my job is to construct the 'notebook' comments that appear both in the official formbook and in *Raceform Interactive* and myself and my colleagues will refer to the route taken by a particular horse if it is of some significance, but neither we nor the race-readers can describe the exact position of every horses at every stage of the race in terms of the route taken, and if you were going to base some sort of system on the position each horse takes in relation to the inside rail then you would need to know all of this. Unless you are able to watch every single All-Weather race and make notes as you go about each horse, this will be an impossible task. Even if you did have the time and the patience to do this, would it be of any real value?

In his book *Beyer On Speed (Houghton Mifflin, New York 1993)* Andy Beyer describes a method used by "trip handicappers" to establish just where a horse was in relation to the inside rail at various stages of the race. They do this by basically allocating 'lanes' just as in athletics. The horse against the inside rail at any stage of the contest is in lane one, the horse on its outside in lane two and so on. So if a horse was racing five-wide at any stage he would be shown to be racing in lane five, and these trip handicappers use their own notations and abbreviations to record these facts.

Using Beyer's own methodology, if he looked back at his notes for a horse's last run and saw this -

3T 5E

then he would immediately know that the horse was three-wide on the clubhouse turn (away from the stands) and five-wide on the bend entering the stretch.

He might use the following for some races –

4FT

which means that the horse was four-wide on the first turn of a longer race.

He also mentioned that it is common practice to adjust the speed figure of a horse depending on its route, such as adding on six points if a horse has gone four-wide around both bends in a two-turn contest, because the horse has lost six lengths in total.

I feel uncomfortable about adjusting speed figures like this, just as I'm uncomfortable about allocating a rating to a chaser or hurdler that has had the race won before falling at the last. A similar scenario occurs with horses that have won eased down, the temptation is to artificially upgrade the figure accordingly in the belief that they could have won by further had they wanted to. This may be right, but there are too many assumptions involved and I prefer to allocate speed ratings on the basis of the time they actually recorded from start to finish, rather than what they might have done. I don't have a problem with adding a symbol such as a '+' after the rating of an eased down winner or a horse that has met trouble in running.

Andy Beyer's way of recording the "trip" or route that a horse has taken looks sound enough to me, and he is one of those people who watches recordings of as many races as possible in order to allocate the notations to each horse as shown above. Reading on a little further in his book, however, he is as uneasy as I am about artificially adjusting ratings to account for what he calls "ground loss" and his reasons for being so cynical sound a resonance with me.

As he points out, going four wide on one track is a disaster whilst going six wide at another may mean very little. It may even be an advantage, as it depends on so much more than purely distance travelled. For instance, when the old Fibresand was in existence at Wolverhampton between 1993 and 2004 it was obvious to anyone who watched racing there regularly that the inside of the track was a no-go area, but that didn't stop many jockeys trying it and becoming unstuck (or physically stuck in the quagmire). I can understand a rider making a beeline for the inside rail on a horse that was starting to struggle, because the temptation to cut the corner and save ground must have been overwhelming. However, it rarely worked unless the weather conditions were extreme and harrowing had evened out the bias, but it usually ended in tears.

The camber on the bends at Wolverhampton meant that the Fibresand rolled down towards the inside rail, making it deeper and harder to get through, and you just knew as soon as your fancy edged towards the inside rail that you might as well start ripping up your ticket there and then.

Had anyone made adjustments to the speed figures of those horses that had raced wide around the bends at Wolverhampton in the belief that they had covered more ground they would actually have been boosting the figures of the very horses that had benefited from a track bias, and penalising those that had raced on the worst ground. If anything, the whole system would have needed to have been turned on its head so that those who raced wide had no adjustments made to their figures whilst those who raced on the inside were bumped up a point or two, but as I have said I don't like messing around with speed figures in this way.

However, another point that Andy Beyer does make is that he will also record a horse's position on the straight parts of the track, not just on the bends. Considering that the notations previously described are a way of accounting for ground loss on the bends, you might ask what is the significance of noting where a horse was on the straight bits? The reasons become very clear when you start to think about track biases.

The way he shows where a horse was on the backstretch (or back straight) might be like this –

2B Rail S

which means that he was two-wide in the back straight and tight against the rail in the stretch (home straight). He does this because it will be of great significance if a track bias exists. For instance how would the performance of this horse look if the inside rail was proving faster than the rest of the track at this particular meeting. If he had won, then there would be grounds for downgrading the effort as he had benefited from a bias. What if the evidence of the whole meeting suggested the rail was "bad" or slower, such as at Wolverhampton in the Fibresand days. If the horse in question had been beaten then there would be a valid excuse. If the animal in question had run a big race in defeat (these were the sort of horses that I was always on the lookout for) then they would immediately go into the notebook, just like horses that had run well against a pace bias.

Unfortunately for most of us the chance to note down the position of every horse in every race is unrealistic unless you really have nothing else to do, but that doesn't mean that the whole concept of track biases is irrelevant because it certainly isn't that. As with pace biases, you must be on the lookout for the presence of such a bias once a meeting gets under way. Of course it's not so easy with track biases as it is with pace biases to gain an edge once you have spotted one. After all, unlike with a pace bias where a glance at a horse's form may help you identify an established running style (front-runner, handy sort, hold-up horse, etc) there is nothing in a horse's form which will tell you how far off the rail he will race once the contest is under way, and whether the horse might be helped or hindered by a perceived track bias as a result. The horse's draw will be of some guidance of course, especially in sprints, but once the race has started you are totally at the mercy of the jockey as to where he or she wants to go.

Track biases, like pace biases, can come and go. Sometimes they will last for just one meeting, but sometimes they can last for months, even years. The anti-inside-rail bias at Wolverhampton lasted for several years and a similar situation existed on the Southwell Fibresand too, at least for races run on the round course. Again, in the vast majority of races if you saw your horse coming up the inside rail at Southwell you started to get that deflating feeling. You just had to be content with making a note that your horse "had raced on the slowest part of the track" and hope that you would be able to profit from that knowledge at some point in the future.

In recent years it has been a major handicap to race tight against the inside rail on the Lingfield Polytrack, whether sticking close to it on a front-runner or in trying to make an effort there from off the pace. Again this was worth noting for the future as the horse in question had probably run a much better race than its final placing would suggest.

One area where it was possible to make money because of a track bias was in races over the straight five furlongs at Southwell, the only straight All-Weather course in the British Isles. The unique point about races run on this track is that the horses basically ran to their draws throughout, rarely deviating very far from their lanes unless they were very green or wayward. In other words you had a good idea about which route each horse would take based purely on its draw. You already knew the size of the field, so the horse drawn highest would almost certainly race tight against the stands' rail (especially as the stalls are usually placed there) whilst in a larger field those drawn low would come up the centre to far side of the track.

The point is that the middle of the track has often been faster than elsewhere, whilst the ground has ridden deeper on the flanks (hence the bias against horses that stuck close to the inside rail). The problem for high-drawn horses in five-furlong sprints is that it appeared to be the same situation for those that raced up the stands' rail, to such an extent that several of the jockeys who were drawn high appeared to become aware of this and started to switch their mounts to their left at the earliest opportunity in order to come more down the centre. However, once this phenomenon had been realised there was money to be made and you didn't have to do very much thinking about it either.

One of the great joys of keeping your own draw and pace statistics is that if a significant bias starts to develop you become aware of it pretty quickly. The numbers begin to shout at you so that you then begin to take special notice of races run over a certain trip at a certain track. If it turns out to be a blip, you will ignore it but if the evidence becomes ever greater that a serious bias has developed, then that is the time to wade in.

The bias that started to develop in races run over the straight five furlongs at Southwell really started to catch my eye during 2002. I enter my data manually into my All-Weather database (very old-fashioned I know) but it does keep me in close touch with what is happening and without doing it this way I would never have been alerted to this particular phenomenon.

I started to notice that low-drawn horses were winning a large number of races over the straight five furlongs, and even more relevant was that I seemed to be entering the number '1' as the winning stall more often than I would usually expect. I started to watch horses starting from that stall with special interest and what I found was that they were winning much more often than they were entitled to. Here are the statistics for all horses starting from stall one over five furlongs on the Southwell Fibresand between 2002 and 2006.

Year	Runners	Winners	%	Profit/Loss (£1 level stake)
2002	37	9	24.3	+62.25
2003	38	4	10.5	-9.09
2004	51	8	15.7	+3.23
2005	56	12	21.4	+44.58
2006	50	11	27.0	+27.75
Totals	**232**	**44**	**19.0**	**+128.72**

The most notable aspect to all of this was not so much the admittedly impressive strike rate of around one winner in every five races, but more the starting prices of the winners themselves. Ten of the 44 winners started at double-figure prices, including three at 20-1 plus a 25-1 shot. This goes some way to explaining the very impressive level-stake profit.

It doesn't seem a comfortable thing to do to back horses 'blind' based purely on their stall number, though it's hard to argue with these statistics. It's also worth noting that for most of the period shown above, according to my pace statistics front-runners were winning around three times more often than they should have, so that combining the statistics (only backing horses starting from stall one with proven early pace) would have boosted the profits even more.

The bias began to ebb away after 2006 and if you had backed stall one blind in 2007 you would have suffered a £17.38 loss to a £1 level stake, so then was the time to draw stumps. This still demonstrates the value in keeping an eye open for a track bias to develop, which they undoubtedly will, and capitalising on them when they do.

That's all very well for races run on a straight course where you can be fairly sure of where the horses will go, but what about in races around a bend? As I said earlier, you are completely at the mercy of jockeys in this situation and it's no good assuming that a horse will race tight against the inside rail simply because it is starting from the highest stall on a right-handed track or from stall one on a left-handed track. So what do you if you have identified a track bias, such as the one that prevailed on the old Wolverhampton Fibresand all those years ago?

The answer is to try and take advantage of something that wasn't so widely available then, but is now – betting in running. This is the only way you can possibly capitalise on spotting a track bias once the race is under way. For instance, if you believe that the rail at a certain track on a certain day is particularly bad, and during the course of a race it looks as though the leader will stick there, you can lay him/her if you wish, or if you spot that a horse is angling for an effort and will be making its run on the part of the track you know is currently favoured, you can back him.

Here is an example of a meeting where there was a significant track bias. It took place at Kempton on January 11th 2008.

The fixture started off with a Class 7 classified apprentice stakes over six furlongs, hardly the sort of race that would get most people out of bed first thing in the morning, but as it turned out this one was to prove significant for the rest of this particular evening meeting, especially to anyone taking notice of what was happening.

The red-hot favourite was a horse by the name of Calloff The Search, who was sent off at odds of 4-6 on account of a victory in a slightly better race at Wolverhampton seven days earlier. He made every yard of the running then, and it seemed that most punters expected him to make every yard now. As soon as the stalls opened it was obvious that the same tactics would be utilized as jockey Jack Mitchell sent the favourite straight to the front and his supporters must have been very happy with the way things were unfolding as the contest progressed.

At around halfway, Calloff The Search was trading at around 1.5 (or 1-2) in running on Betfair, whilst at the back of the field following a tardy start one of the outsiders Stoneacre Donny (SP 20-1) was available at 50-1. However, things started to get interesting once the field straightened up for home (they were using the outer loop for this contest) as Stoneacre Donny suddenly got a dream run tight against the inside rail and was now right on the favourite's heels. As he did so, his in-running price tumbled right down to 7 (or 6-1) and then when his rider Ryan Hill made a beeline for the inside rail after the intersection where the two tracks meet to get up the inside of Calloff The Search, his price dropped to 6. A couple of seconds later he was right on terms with the favourite tight against the rail and his price dropped to 4.2.

The remarkable thing was that whilst all of this was going on, the favourite's price was dropping too! The only reason I can think of was that the contest was now effectively just a two-horse race. Even when Stoneacre Donny was down to 4.2, Calloff The Search fell to a very skinny 1.18 even though it would have been obvious to anyone watching that we were about to have a new leader. It wasn't long before the prices of the two horses flip-flopped, with Calloff The Search drifting out to 2 (or evens) whilst Stoneacre Donny was in to 1.55.

After a brief struggle, Stoneacre Donny powered clear to beat the favourite by three lengths with another six lengths back to the others. Those that had gone in short on the favourite in running had burnt their fingers, but more significantly was this the first sign of a bias towards the inside rail?

The second race on the card was another Class 7 classified event, though this time over seven furlongs.

The 5-2 favourite was Sion Hill, who attempted the same forcing tactics as Calloff The Search had in the opener, though in his case he was ultimately successful. In fact this contest was very straightforward, as the favourite pinged the gates and never looked in much danger at any stage. He was briefly available at 3.8 after a few strides, but for the majority of the race he traded at much lower than his SP, as basically the hardest part (gaining an uncontested lead) had already been achieved.

However, the most significant apart of the race with regards to betting in running was again passing the cutaway at which point his jockey Mickey Fenton took him right over to the inside rail and his price quickly dropped from 1.23 to 1.08. It continued to fall after that, which was understandable as he eventually bolted up to win by three-and-a-half lengths. His performance didn't really prove the existence of a bias because he was the best horse in the field and he had the run of the race, but it didn't disprove it either.

The next race, a six-furlong handicap, would also prove to be an interesting contest, though in a slightly different way.

The early leader was Regal Royale, a stable companion of the earlier winner Stoneacre Donny, and those who had seen Sion Hill make all in the previous contest were obviously expecting a repeat, because even before the field had reached the cutaway Regal Royale had plunged right down to 1.15 in running. However, anyone who had backed him and suspected the presence of a rail bias and therefore expected him to be manoeuvred over to the inside rail after the cutaway was to be sorely disappointed. Despite no other horses racing on his inside, he was kept in a straight line which meant that he was now a couple of horse-widths away from the inside rail instead of tight against it.

Even so, he continued to trade at odds-on for a good ten seconds afterwards until the 2-1 favourite Mind Alert was produced to mow him down late. Admittedly the winner came down his outside, but it still would have been fascinating to see if it would have been a closer finish had Regal Royale bagged the far rail when he had the opportunity.

The next race on the card was a one-mile-three-furlong claimer and this contest demonstrates perfectly what can be achieved if you have made a decision about how a track is riding and are willing to invest money on that decision.

The majority of the running was made by the second-favourite Sweet World, who still held a clear lead on the crown of the home bend and was matched at a low of 1.9. However, as with Regal Royale in the preceding race, his rider didn't make use of the cutaway and stayed straight, leaving an inviting gap on his inside for anyone who wanted to use it. Meanwhile, Neil Pollard aboard the outsider Steely Dan, who had been switched right off out the back early, had been sneaking his way into contention tight against the inside rail running towards the intersection (very much like Stoneacre Donny had done in the opener) and he was in the ideal position to dive to the inside if the leader decided to carry on straight. Anyone who had anticipated what Neil Pollard was going to do could have got on him at an in-running price of 32, but they would have to have to have been quick because as soon as Pollard made his move, his price collapsed to 2.5 (or 6-4). He managed to see his race out well enough from that point to leave his backers happy and give further credence to the presence of a rail bias.

The fifth race on the card was a one-mile handicap in which Simpsons Gamble went off the 7-2 favourite despite being a maiden after 22 attempts. The early running was made by the 25-1 outsider Golden Square and by the time the field started up the home straight enough people had become interested in him to drive his in-running price down to 8. It was then that Simpsons Gamble was delivered up the leader's inside and it was he who managed to bag the inside rail after the cutaway. At that stage Golden Square's in-running price had shot right out to 95 and under normal circumstances that would have been that, but Simpsons Gamble wasn't a long-standing maiden for nothing and waived the opportunity to break his duck. That meant that Golden Square was now back in front and his price dropped right down to 3.85. It also meant that he was in a better position one off the rail than those trying to launch their efforts down the wide outside, and he managed to hold them off.

The penultimate contest was another one-mile handicap, though a rather better-quality one.

The 5-2 favourite was Alfresco, but it was one of the 7-2 joint-second favourites Monkey Glas, an established trailblazer, that set the pace. His rider Andrew Elliott, who had declined to bag the inside rail aboard Sweet World after the cutaway in the earlier claimer, eventually took Monkey Glas over there this time and that may have won him the race. It did appear that he was a sitting duck for Alfresco once Richard Hughes delivered the favourite with his effort between horses, and Monkey Glas's in-running odds drifted right out to 14 whilst Alfresco's collapsed, reaching a low of 1.03, but remember that Monkey Glas had the rail and he found reserves from somewhere to snatch the prize by the narrowest of margins.

The final contest was a modest handicap over a mile and a half. The short-priced 6-5 favourite was Smokey The Bear who was held up right out the back early, but that didn't appear to bother anyone as his price in running hardly veered far away from his SP, reaching just over 3 (or 2-1) on a couple of occasions. The one that seemed to be in the ideal position on the inside just behind the leader Under Fire was Richard Hughes's mount Little Richard, but despite that his price in running hit a high of 18.5, even though he was travelling sweetly throughout.

Smokey The Bear made his move as soon as the field straightened up for home and his in-running odds tumbled accordingly, though it was likely that he was going to be forced to make his effort wide. For over 20 seconds (or nearly two furlongs) he was odds-on, reaching a low of 1.19 and indeed he made his effort wide of Under Fire. Meanwhile Little Richard, given a fine tactical ride by Hughes, dived for the inside after the cutaway but was still available at 12 at that point (admittedly only to small money) and after dropping right down, his price went back up to 12 when it looked as though Smokey The Bear would swamp the lot of them. However, just as Monkey Glas had done half an hour earlier, Little Richard kept on finding a bit more (possibly aided by a faster strip of ground against the rail) and battled on to prevail in a three-way photo from Under Fire and Smokey The Bear.

The bias continued for a few more meetings afterwards, therefore increasing the potential for profits for those in the know. These sorts of biases happen all the time and it's just a case of latching on to one as soon as possible, but only when you are sure enough that one does exist.

The January 11th Kempton meeting also helps demonstrate the sort of in-running opportunities which now exist for those who keep their eyes open and look for biases on the All-Weather tracks, whether a pace bias or a track bias. It is also a useful opportunity if you know that certain parts of a track are at an advantage or a disadvantage because your historical data tells you so. It's an advantage we didn't used to have and one it would be a shame to waste.

SECTIONAL TIMES

THE HEADLINE of the article within the November 12th 2007 edition of the *Racing Post* said it all:

"All-Weather results in the Post will come with added sectionals."

The article began with:

"The Racing Post *marks the start of the All-Weather season with another stage in the continual evolution of its data provision today. Sectional times will now be included in the result of All-Weather races covered by TurfTrax."*

It seemed that now, finally, British racing had started to catch up with the rest of the world and make sectional times an integral part of everyday racing results. In the United States for instance, the idea of not having sectional times available to bettors would be like not being able to find a bible in a church. The only concern was whether people would know how to use them, not that it should have been. Rather than assuming that the majority of people who follow racing closely and/or like to have a bet are a complete bunch of dummies, better to give them the data and let them decide how they want to interpret it. You can advise or demonstrate a method of doing it of course, which is what this chapter is for, but at the end of the day people have the power to not use available data or to use it in their own way, however they see fit. Contrast the earlier *Racing Post* headline with the one that greeted its readers in the July 4th 2008 edition:

"TurfTrax pulls the plug on sectional timing."

The opening part of the piece made for depressing reading and starkly put everything into context:

"Sectional Timing on British racecourses is a thing of the past, following a radical company restructuring by its provider TurfTrax ..."

This exciting innovation had lasted just eight months and such a short period of time can only be described as shameful. Of course there were many people who were cynical about the whole project when it started, fearing that there was likely to be too much information for the everyday punter to digest, or that the different conformation of British tracks would make sectional timing meaningless, but that suggests a lack of will to actually sit down and think about it and come up with a solution.

The way that a horse's form is shown generally in the racing media is incredibly complicated if you are not familiar with it and different publications have very different ways of showing the same thing, but I'm sure that no-one would advocate that form shouldn't be published anywhere because Joe Public might find it all too confusing. As far as the wide variety of British tracks is concerned, that hasn't stopped speed figures being compiled over here for decades so it must be possible to overcome that obstacle.

Perhaps when the whole sectional timing episode eventually fell in a heap in July 2008 there may not have been much evidence that it had made much difference to people's betting habits, but we just don't know because there wasn't enough time to find out. Sadly that is the situation at the time of writing, but there is always the chance that it has changed again by the time you are reading this. I certainly hope so.

The decision I had to make whilst in the middle of writing this book was whether to delete this chapter completely together with all the research I had already carried out in order to help to create it, but in reality there was no decision to be made at all. My belief is that what we saw within that eight-month period from November 2007 to July 2008 was of significant enough value to make it a subject which really should be covered. Surely one day someone will pick up the baton and run with it so that sectional times can once again become part of everyday form study for those who want to use it. So this is how I used sectional times as part of my All-Weather racing database and how I will use them again when they return.

Back in 1998 Newmarket were in the process of carrying out some trials with sectional timing. I was asked by the managers at *Raceform* to think of a way that they could be interpreted for use in the official formbook. The idea I had at the time, which is hardly an original one, was to come up with a series of 'standard split times', i.e. the average time it took for the race leader to reach each split (points at various stages during the race where the split and cumulative times were recorded) over a series of races over each distance. The idea never really developed fully then, but it all came back to me when TurfTrax and the *Racing Post* got together to make sectional times widely available for All-Weather meetings within the paper's results section in the autumn of 2007.

Admittedly the data provided by TurfTrax did have holes in it. Not every All-Weather meeting was covered and even those that were often had one or two races missing, but I put up with that in the belief that those holes would be filled, and anyway I was still trying to work out a way of incorporating the information provided into my database before attempting the necessary programming in order to make it happen.

The way that the sectional times were shown in the *Racing Post* was something like for this hypothetical seven-race meeting –

3.00	**5f**	**2f**	**4f**	**WP**
	Cumulative	25.20	47.50	59.70
	Individual	25.20	22.30	12.20

4.30	**5f**	**2f**	**4f**	**WP**
	Cumulative	25.60	47.80	60.10
	Individual	25.60	22.20	12.30

2.00	**6f**	**2f**	**4f**	**5f**	**WP**
	Cumulative	24.50	46.90	59.40	1:12.20
	Individual	24.50	22.40	12.50	12.80

3.30	**7f**	**2f**	**4f**	**6f**	**WP**
	Cumulative	25.80	50.30	1:13.40	1:25.40
	Individual	25.80	24.50	23.10	12.00

2.30	**1m**	**2f**	**4f**	**6f**	**7f**	**WP**
	Cumulative	26.40	51.60	1:15.80	1:28.20	1:41.10
	Individual	26.40	25.20	24.20	12.40	12.90

5.00	**1m**	**2f**	**4f**	**6f**	**7f**	**WP**
	Cumulative	27.10	52.20	1:16.20	1:28.40	1:41.20
	Individual	27.10	25.10	24.00	12.20	12.80

4.00	**1m5f**	**4f**	**6f**	**8f**	**10f**	**12f**	**WP**
	Cumulative	49.80	1:16.00	1:41.10	2:06.80	2:32.30	2:45.20
	Individual	49.80	26.20	25.10	25.70	25.50	12.90

The 'cumulative' time was how long it had taken to reach that point from the start of the race, whilst the 'individual' time is how long it had taken from the preceding point at which the elapsed time had been recorded. In most cases the time taken to reach the first split was the slowest because the horses had begun from a standing start.

I have to admit that there are quite a few figures shown here and utilising them all would be a major, though not impossible task. My view is, however, that the most significant part of the race in terms of pace is the first couple of furlongs. Sometimes the field can go slow early before quickening up after a quarter of a mile or so, but in most cases the pace or tempo of the race is determined in the early stages. If they go quickly early, then more likely than not the contest will be a truly run one, whilst if they go slowly early the race is likely to develop into a sprint. In my experience the tempo of All-Weather races is much more uniform and consistent than it is on turf. In other words if the runners go fast early on sand, the second half of the contest will be relatively slow and vice versa.

Because of the nature of the sand surfaces you get fewer races run at a stop/start gallop than you do on turf, so if you want to get some sort of a guide as to how truly run a racc on sand was, the sectional time to focus on is the first one, because as I have said before the tempo of the contest will have been determined within that section of the race. It also means that I can ignore all the sectional times apart from those in the first column shown above. I'm not saying that you should. By all means use as many of them as you wish, but

I'm content that the data I use will give an interesting insight into the shape of a particular race. By definition that will also tell me whether an event has been truly run or not, and consequently whether the form is reliable or if a horse's stamina has been truly proven or not.

Having reduced this apparently large amount of data down to manageable proportions, how do you go about using it?

The first thing to notice about the way the sectional times have been displayed above is that the races are not shown in chronological order, and there is a very good reason for this. By ordering them in terms of race distance, i.e. shortest at the top, longest at the bottom, it means that any two or more races which were over the same distance will appear next to each other, and therefore make comparisons between cumulative times, or individual splits, much easier to spot. For instance, at the above meeting the top two races were both run over five furlongs and it can be seen that the leader in the first race reached the first split (or call) in a time of 25.2 seconds, whereas in the second contest it took 25.6 seconds, or 0.4 seconds slower. That equates to about two lengths. Moving on through the two races, the second individual split in the second race was 0.1 seconds quicker than the first, but in the earlier contest the final split was 0.1 faster. That meant that the final time for the 3.00 race was 0.4 seconds faster than the 4.30.

Looking further down the meeting, you will notice that there were two races run over one mile and the initial split for the 2.30 contest was 0.7 seconds quicker than for the 5.00, which is about three-and-a-half lengths. Admittedly the winning times for each contest were only 1/10th of a second apart, meaning that the runners in the 5.00 race had put in the bulk of their effort in the later stages of the contest. The winner of the 2.30 race had still recorded the slightly faster overall time (though on my scale they would probably have received the same speed figure) but the 2.30 contest had been run at a much more consistent end-to-end gallop and that fact ought to be recorded somewhere.

Even that information as it stands is of some value, as you already know more about these four races than the vast majority of punters would. You could just leave it at that, and record these details somewhere in your formbooks, on your computer, etc. However, there is an opportunity to make this data even more meaningful. For example, with the two five-furlong races we know that in one of them the leader covered the first two furlongs in a time 0.4 seconds faster than the other, but how do these two split times compare with the split times of other races over the same trip on other days? Was one of them good and one of them bad? Were they both good or bad? Unless you have some sort of benchmark to compare split times with, you will be merely guessing.

The way I went about establishing these benchmarks was to record each first split time for every distance at each All-Weather track over a certain period of time, omitting races confined to two-year-olds. If this all sounds very familiar it should be. The method I used was precisely the same as the one used to establish the median times in Chapter Seven, though this time there is no need to bother with race classes, pars, etc. As with the median times, once a sufficient number of races had taken place so that the sample for each race distance was large enough (hopefully with an odd number of entries) I ended up with a listing like this once the split times had been sorted into ascending order –

5f	6f
24.2	24.3
24.2	24.4
24.4	24.5
24.5	24.5
24.5	24.6
24.7	24.8
24.8 *	25.0 *
25.0	25.1
25.2	25.3
25.3	25.4
25.3	25.6
25.6	25.9
25.7	26.1

Note: Sectional Times were never available at either Great Leighs or Dundalk.

The median split times for each of these two distances are indicated by the asterisks and the first split time for every subsequent race over the same course and distance will be compared with them.

Once I had a full set of median splits for every distance, the decision was then how to use them to maximum advantage. As I said earlier, one possibility would be to just do a straight comparison between the first split of a particular race and the standard split for the distance. After all, one thing we now know is that split times of 25.2 seconds and 25.6 seconds for the two five-furlong races at our hypothetical meeting are both slower than average if we apply the median splits shown above. There could be any number of reasons for this, perhaps the early pace was modest or the track was riding slow, but either way this is something we wouldn't have known without some sort of benchmark. Incidentally the split time for the six-furlong race (2.00) was half a second faster than average using the above median.

That still wasn't really enough for me though. I wanted some sort of figure that I could include within my all-in-a-line All-Weather form which would tell me instantly whether the race was run fast or slow early. So this is how I went about it.

Taking the two five-furlong races, the first split for the 3.00 contest was 0.4 seconds slower than the median split (25.2 – 24.8) whilst the 4.30 race was 0.8 seconds slower (25.6 – 24.8). In order to convert these figures into points (or lengths as per the speed figures) they need to be divided by the amount of time it takes to cover a length, which is around 0.2 of a second. Doing so for the 3.00 race would give (0.4 / 0.2) or two lengths slower than the median split, whilst the 4.30 would be (0.8 / 0.2) or four lengths slower than the median split.

You could leave it at that, but as with the median times used in the calculation of speed figures there is a problem. On average the margins between split times and the median split increase the longer the race, so once again everything is standardised so that it appears that all races have been run over one mile. As the first split for the examples races comes after two furlongs, converting everything to one mile means multiplying the difference between the split time and the median split by four. In the 3.00 race that means multiplying 0.4 by four, giving 1.6, and then dividing that by 0.2 (or multiplying by five gives the same thing) which produces a result of eight lengths per mile slow. In other words, had the runners in the 3.00 race maintained the same velocity up to a mile they would have been eight lengths slower than the median split. In the 4.30 race the result would be 16 lengths slower than the median split (or +16).

It therefore follows that the 'sectional rating' as I call it for the 2.00 race over six furlongs would have been –10, or ten lengths per mile faster than the median split time.

Note: the first split for races beyond a mile and a half came after four furlongs rather than after two, so in order to standardise the differences between the split time and the median split to one mile over those trips means multiplying the difference by two rather than four.

Having reached that stage, I was then faced with a dilemma and it is a purely personal choice as to how you go about tackling it. What do we do about the state of the track? It's not an easy question to sort out. Whilst the going will have a major impact on the overall time of a race how much effect it will have on the first couple of furlongs is harder to gauge. It seems to me that if the surface is very testing that will have a bigger impact on the horses towards the end of a race rather than at the beginning. Often they fly out of the stalls and go hell for leather over the first two furlongs and it is only when they start to tire that the surface starts to get to them.

I'm sure that calculating sectional ratings the way I have described thus far gives an invaluable insight into the shape of a race, but I find it very hard to ignore the state of the ground in any calculation so I decided to make an adjustment for the going. The good news is that very little extra work was involved, because I already had a good idea as to what the state of the ground was, I had already calculated a track variant as part of the speed-figure process and it was expressed in terms of lengths per mile, just as the sectional rating is. So, for instance, if I had worked out that the track variant for our hypothetical meeting was +4 (or slowing the horses down by four lengths for every mile they ran) then the sectional ratings would be adjusted like this.

The original sectional rating for the 3.00 race was eight lengths per mile slow, but four of those lengths were on account of the slow ground, so the adjusted sectional rating for this race would be +4 (+8 – 4). The adjusted sectional rating for the 4.30 race would be +12 (+16 – 4), whilst for the 2.00 race it would be a very impressive –14 (-10 – 4). This rating for the 2.00 contest is just the sort of sectional rating I want to know about. I have evidence at my fingertips that this race was run at a serious pace early and if the final time was decent too then this was a strongly run race throughout and almost certainly rock-solid form. If, on the other hand, the final time was modest then there is a probable valid reason for it, i.e. they went off very hard before the pace collapsed.

If the winning time was good for either one or both of the five-furlong races despite modest first splits, then I can be confident that the latter part of each contest was run at a stronger pace in order to make up the time. In other words a combination of the sectional rating and the speed rating of a particular race will tell me everything I need to know about it in terms of distribution of effort. The racing media does a fine job at trying to describe the early pace of a race and if a contest is particularly slowly or strongly run then an experienced race reader will put you on to it, but there is nothing quite like having hard facts at your disposal to do with as you please, and the data that TurfTrax provided did just that.

So now I have these sectional ratings incorporated within my All-Weather form, like this –

DANSANT											
03 NOV 2007	KEMP	12	FT	1LST	1	8	8-11	2	H	---	111
01 DEC 2007	KEMP	12	SD	1LST	1	8	9-2	6	H	---	112
23 FEB 2008	LING	10	SD	1LST	1	12	9-2	4	H	-16	107

The sectional rating is in the second column from the right, and in this case it tells me that in the race won by Dansant at Lingfield on February 23rd, the field covered the first two furlongs 16 lengths per mile faster than the median split time having allowed for the conditions.

A slow sectional rating doesn't necessary mean the race was bad, especially if the winner's speed rating is good, but it may mean that one or two horses in the race might have been inconvenienced, especially those whose stamina was their forte and who were done for finishing speed because of slow early fractions.

Talking of slow early fractions, I had to do quite a bit of programming in order to incorporate the sectional time data into my database and it provided a painful lesson. One decision I had to make was how many characters I would allow for this particular variable. Including the leading '+' or '-', I believed that three characters would suffice as you would never get a race where the runners would register an initial split time over 99 lengths slower than the median, and they would certainly never cover it more than 99 lengths faster! I sat back with a smug feeling of achievement once the system was up and working; the sort of smugness that often ends in tears.

Everything was working fine until a four-runner race took place over the extended nine furlongs at Wolverhampton on 27th March 2008. As soon as I saw the race I knew I was in trouble, they went incredibly slowly early and even though things eventually quickened up, the winning time was still very slow. It was with a real sense of trepidation that I processed the result on my home computer the following day.

The sectional rating of the race came out at +5, which surprised me a bit as that was nothing like as bad as I had feared, but just to be on the safe side I decided to check the mathematics with a calculator. Lo and behold the actual sectional rating came out at +105! My system had chopped off the '1' because it was beyond the range of characters I had allowed and then chopped off the leading zero because it wasn't needed. Somehow, this field had run the first two furlongs of their race in a time 105 lengths per mile slower than the median split allowing for the ground. They had also caused me to spend several more hours rewriting the computer program again, though I don't suppose the four jockeys involved lost any sleep at all over that.

One area where I have found this methodology especially valuable is with staying races. By their very nature staying races are often slowly run in the early stages, so that when one returns a fast sectional rating it tends to stick out. Take for example the two-mile handicap won by Rollin 'n Tumblin at Lingfield on March 5th 2008. Despite there only being five runners which would often mean a steady early pace, this time there was no hanging about as the eventual third Dash Of Grey set serious early fractions, therefore making sure that the race would become a true test of stamina. The winner's speed figure was good too, so it was no surprise that the form stood up. In fact all five horses were successful within their next three starts.

Since sectional times disappeared from the picture in July 2008, I have missed them and my All-Weather racecards look rather bare without them. Hopefully this chapter will have shown that they can have some value if you can figure out a way of using them, and that they will be back in the public domain sooner rather than later.

THE SHAPE OF THE RACE

THIS IS an area of horse racing, not just on sand, that tends to be ignored by most punters, so therefore remains one of the few areas that can be exploited and provide an edge. The reason why this aspect is still basically ignored by so many is mainly down to laziness. After all, it's all very well scanning your racing paper, or the racing pages of your daily paper, picking your horse by whatever methods suit you, and putting your money down. The problem is that whilst the horse that you consider possesses the best credentials might be able to show its true form and justify the faith you have placed in it, there will be occasions when it is unable to do so, not always through any fault of its own and quite often because of other runners in the race. I don't mean that your selection is beaten by these horses, but because they stop yours from running its race.

A horse's running style is hugely important. So many are only able to run up to their best when they have the race run to suit and this is especially true of front-runners. Horses like these are completely transformed when they can get their own way out in front, either because they enjoy bullying the opposition into submission or more often because they don't like being surrounded, especially in the early stages of a race when the attentions of other horses cause the established trailblazer into expending nervous energy, which will eventually count against them as the race progresses.

So where can you find out what a horse's normal running style is? One way of course is through your own eyes. If you watch enough racing either live or on TV you will get to know who the habitual pacesetters are, and these days the television racing media are very good at isolating horses with a particularly running style. They will also often give their opinion as to whether this style will be helpful or a hindrance for the contest they are analysing. In other words they have worked out what the likely shape of the race will be.

If you aren't able to watch that much racing, then you will have to rely on the trade paper or a formbook, either paper or electronic, in order to research each horse's running style in its past performances. Within its form section for each meeting due to take place, the *Racing Post* will show the comment-in-running for each engaged horse's last few starts and this will at least give you an idea of how the horse has been ridden lately. Even this fairly limited data is better than nothing and looking at it will already give you an advantage over most punters.

I may be biased because they are the company I work for, but *Raceform* does some of the work for you in many of their publications by allocating a pace symbol to each horse within its racecards. They are very similar to my own symbols of (L)ed, (P)rominent and (H)eld Up although their symbols are a general description of the horse's running style based on its past performances, whereas mine are on a race-by-race basis.

In his book *Betting For A Living (Aesculus Press, 1992)* Nick Mordin advocates a 'speed points' system based on each horse's last three starts, itself based on the speed points system described by Professor William Quirin in his book *Winning At The Races – Computer Discoveries in Thoroughbred Handicapping (William Morrow, New York 1979)*. Basically taking each of those last three outings in turn, a horse is allocated four points if it led before halfway, two points if it disputed the lead before halfway, and one point if it was prominent before halfway. If none of these scenarios prevailed then no points were allocated, so it can be seen that the maximum score a horse can achieve in three outings in terms of speed points is 12, whilst the minimum is zero.

I used this system myself both on sand and on turf for a few years and it worked very well, at least in helping me work out who the early pacesetters were and how many of them there were likely to be in one race. I found it especially fascinating in races contested by a large number of runners where the fields traditionally split, such as the Wokingham, Stewards' Cup, Ayr Gold Cup, etc. By writing the stall numbers in descending order down a sheet of A4 and recording the 'speed points' for the occupant of each, I had a good idea as to where the trailblazers were drawn and as a result whether one side was likely to dominate the other in the early part of the race, and very possibly throughout it. I would still be using this methodology now had it not been for the fact that I started to create my own computerised All-Weather database in the mid 1990s and was able to record 'early speed' data in a way that suited that system.

The way I do it now is to record the position each horse took within the early stages of each of its races using the symbols previously described (L, P and H). So if the comment-in-running for the individual concerned was any of the following – led, soon led, made all, made most, etc, the horse would be allocated the 'L' symbol for that race. If the comment was anything like – prominent, close up, chased leaders, tracked leaders, pressed leaders, etc, then the 'P' symbol would be used. The 'H' symbol is used for anything else. You could use 'M' for midfield. In fact you can get as detailed as you like, but I'm not really that bothered whether the horse is in midfield, is outpaced or is tailed off. The horses I'm interested in are those whose jockeys have intentionally ridden them up with the pace, as this will be very useful information when the horse runs again and will give me a great idea as to how many horses will be vying for the early lead. In other words it will help me work out the likely shape of the race.

Here is an example of the sort of data I have available to me using the methods just described. The pace symbols for each of this horse's past performances on sand are in the third column from the right.

THE JAILER

24 FEB 2006	WOLV	8.6	SD	6SEL	5	7	8-3	8	P	---	82
08 MAR 2006	SOUT	8	SD	7CLF	2	10	8-12	2	P	---	90
05 APR 2006	KEMP	10	SW	7CLF	3	10	9-0	8	H	---	92
25 APR 2006	WOLV	8.6	FT	7CLF	2	13	8-12	13	L	---	92
15 MAY 2006	KEMP	8	SD	7CLF	3	14	9-0	3	L	---	84
13 SEP 2007	WOLV	8.6	SD	7CLF	5	13	9-3	4	H	---	93
04 OCT 2007	KEMP	7	FT	6HCP	1	13	8-5	12	L	---	95
03 NOV 2007	KEMP	8	FT	6CLM	9	11	8-6	3	L	---	96
21 NOV 2007	KEMP	7	FT	6HCP	2	13	8-8	8	L	---	100
10 JAN 2008	WOLV	7.1	VS	6SEL	7	7	8-3	4	L	---	90
02 FEB 2008	LING	8	SD	6SEL	2	3	9-0	1	L	+9	98
15 FEB 2008	LING	7	SD	6HCP	2	8	8-13	4	L	+7	98
02 APR 2008	KEMP	7	SW	6CLM	4	8	8-2	7	L	-14	99
10 APR 2008	KEMP	6	SD	5HCP	3	5	7-12	6	L	+16	101
28 APR 2008	LING	7	SW	6SEL	3	10	8-9	2	L	+3	99
09 JUL 2008	KEMP	7	SD	5HCP	5	12	8-11	5	L	---	96

Form shown in this way will leave little doubt that The Jailer is an established trailblazer.

I'm not saying that this will enable me to predict whether The Jailer is going to win or not, there are still many other factors to consider. What I can say is that if there aren't any other horses with a similar profile to her about to take her on, then she will have a much better chance of giving it her best shot. If there are one or more similar types to her in opposition, then I would very likely swerve the lot of them and look elsewhere, probably for a horse that will be suited by getting a tow whilst a complete burn-up takes place up front. Here is an example of what I have been talking about..

The evening meeting at Wolverhampton on Thursday January 10th was a pretty typical fixture of its type, nothing great standard-wise, but as I've said before who cares when you can make some money? The fifth race on the card was a four-runner fillies' handicap over the extended nine furlongs. One of the quartet was a five-year-old mare named Chia, trained by Derek Haydn Jones and the mount of Hayley Turner. Admittedly Chia's profile wasn't that convincing at first glance, as she had raced a total of 35 times prior to this and won only once, though that was over this course and distance almost a year earlier.

The notebook comments in the official formbook for some of her recent outings had a similar theme running through them, i.e. *"is proving very hard to win with"*, etc. Chia liked to force the pace when she could as can be seen from her All-Weather record as it appeared in my database.

05 DEC 2005	LING	7	FT	5HCP	6	13	8-10	11	H	---	95
27 DEC 2005	SOUT	8	SD	4HCP	3	9	8-12	2	L	---	95
29 JAN 2006	LING	8	FT	5HCP	7	12	9-0	3	P	---	90
02 MAR 2006	LING	8	FT	5HCP	5	9	8-12	1	P	---	99
24 MAR 2006	LING	8	SD	5HCP	4	11	8-13	1	H	---	98
30 AUG 2006	WOLV	8.6	VF	5HCP	9	11	8-9	4	H	---	95
01 NOV 2006	KEMP	8	FT	6HCP	2	12	8-13	6	P	---	94
11 NOV 2006	WOLV	8.6	FT	6HCP	2	12	8-13	5	P	---	97
28 NOV 2006	WOLV	9.5	SD	5HCP	3	9	8-9	2	H	---	104
18 DEC 2006	WOLV	8.6	SW	5MDN	2	12	8-12	7	P	---	98
10 JAN 2007	KEMP	8	SW	6HCP	2	14	9-0	6	L	---	96
29 JAN 2007	WOLV	9.5	SD	5HCP	1	9	8-10	1	P	---	101
10 FEB 2007	WOLV	8.6	SD	5HCP	6	8	8-7	1	H	---	95
05 APR 2007	WOLV	8.6	SW	5HCP	2	7	9-0	2	L	---	104
10 JUL 2007	WOLV	12.2	SD	5HCP	4	11	9-9	7	P	---	94
31 AUG 2007	WOLV	12.2	SD	5HCP	4	10	9-11	5	P	---	103
26 SEP 2007	KEMP	12	FT	5HCP	6	13	9-0	10	P	---	100
18 OCT 2007	WOLV	9.5	SD	5HCP	3	12	8-11	4	P	---	99
08 NOV 2007	WOLV	9.5	SW	5HCP	2	9	8-10	2	L	---	103
17 NOV 2007	WOLV	9.5	SW	5HCP	2	9	9-2	8	P	---	103
01 DEC 2007	KEMP	8	SD	5HCP	6	13	8-9	7	H	---	96
26 DEC 2007	WOLV	9.5	VS	5HCP	2	8	8-13	3	L	---	97

The problem for Chia seemed to be that when she did get to the front early, she either had to fight to get there or was taken on from a long way out and never seemed to quite get home. Her form figures on sand when she was held up were 649366, whilst her form figures when leading early were 32222. Here was a wonderful chance surely.

Two of her rivals, Sforzando and Casablanca Minx (more about her later), were confirmed hold-up horses whilst the only other runner, Marias Dream, was having only her second start on sand and she hadn't led early in the first. For fans of the 'bounce' theory, she was also running just five days after running a big race on her return from a five-month absence. Surely Chia was very likely to get an uncontested lead this time, but the punters sent Marias Dream off the 5-6 favourite whilst Chia went off at 7-4. Apart from not winning as often as she should everything else (the shape of the race, the speed figures, etc) made Chia the one to beat. A decent bet just had to be made.

The race could hardly have worked out better for Chia (or me) either. She soon bagged an uncontested lead, set a very modest early pace and then had more than enough energy left to hold off Marias Dream's final effort. Without looking deeper than just the obvious form, this race would have looked nothing like as straightforward.

Chia was out again over the same course and distance seven days later under a penalty. She faced five rivals this time, but there was a major threat lurking amongst them in the shape of a horse called Alonso De Guzman who had run really well when forcing tactics were applied at Wolverhampton the day before Chia's victory. If similar tactics were applied here, then Chia was going to be in trouble. They were, and she was. In fact Alonso De Guzman went on to win whilst Chia dropped away to finish fifth of the six runners. Her defeat was nothing to do with carrying penalties or making a quick return to the track. This was all about the shape of the race being suitable one week, but not the next.

Here is a good example of when having information about running styles was again very relevant. There was a valuable five-furlong handicap due to take place at Lingfield on February 2nd 2008 for which ten runners went to post. Amongst them were two renowned trailblazers, as you can see –

MALTESE FALCON

21 OCT 2006	LING	5	SD	2HCP	4	10	8-9	10	H	---	103
01 NOV 2006	KEMP	6	FT	3HCP	9	11	9-4	1	P	---	103
11 NOV 2006	LING	6	SD	2HCP	1	10	8-4	4	P	---	113
01 DEC 2006	LING	6	SD	3HCP	1	8	9-1	4	L	---	110
16 DEC 2006	LING	6	SD	2HCP	5	9	9-0	1	L	---	111
01 JAN 2007	SOUT	5	SD	2HCP	8	8	9-4	6	P	---	95
20 JAN 2007	LING	6	SD	2HCP	2	10	8-13	5	L	---	111
03 FEB 2007	LING	5	SD	2HCP	2	10	9-2	1	L	---	107
17 FEB 2007	LING	5	SD	2HCP	2	10	9-6	3	L	---	106
24 MAR 2007	LING	5	SD	1LST	4	10	9-0	10	P	---	104
01 APR 2007	LING	6	SD	2HCP	1	4	9-2	1	L	---	104
17 NOV 2007	LING	6	SD	2HCP	4	11	9-9	9	L	---	109
24 NOV 2007	LING	6	FT	1LST	1	9	9-2	8	L	---	108
28 DEC 2007	LING	6	SD	2HCP	4	9	9-9	8	P	---	104

SILVER PRELUDE

23 APR 2006	KEMP	5	SD	4HCP	10	10	9-6	9	H	---	96
11 JUN 2007	WOLV	5.1	SD	6HCP	2	13	9-4	3	L	---	99

10 JUL 2007	WOLV	6	SD	6HCP	2	12	9-3	13	L	---	103
29 AUG 2007	KEMP	5	FT	5HCP	8	12	9-0	3	P	---	94
31 AUG 2007	WOLV	6	SD	6HCP	2	13	9-7	2	L	---	97
06 SEP 2007	WOLV	5.1	SD	6HCP	2	12	9-4	2	L	---	103
20 SEP 2007	WOLV	5.1	SW	6HCP	1	12	9-3	4	L	---	101
21 OCT 2007	WOLV	5.1	SD	5HCP	2	12	9-1	2	L	---	97
02 NOV 2007	LING	5	FT	5HCP	5	9	8-8	10	P	---	99
30 NOV 2007	LING	5	SD	5HCP	1	8	8-9	10	L	---	105
05 DEC 2007	WOLV	5.1	SD	5HCP	1	8	8-10	3	L	---	106
12 JAN 2008	LING	6	SD	3CND	2	7	9-0	8	L	+7	99

There was every chance that these two were going to cut each other's throats, so although they were both good enough to win at their best, they couldn't really be supported. What was needed was a horse that liked to come from behind, was suited by the track and had demonstrated that it was fast enough. A horse like this one would do nicely.

EBRAAM											
21 FEB 2007	WOLV	7.1	SD	6HCP	7	10	9-2	6	H	---	92
15 MAR 2007	SOUT	7	SW	6HCP	13	14	9-3	10	P	---	82
16 APR 2007	WOLV	7.1	SD	6HCP	6	12	9-4	11	H	---	93
24 APR 2007	WOLV	6	SD	5HCP	6	10	8-8	9	P	---	98
29 JUN 2007	WOLV	6	SD	6HCP	4	12	9-4	12	H	---	99
10 JUL 2007	WOLV	6	SD	6HCP	1	12	9-5	4	H	---	103
25 JUL 2007	LING	6	SD	5HCP	2	11	9-5	4	H	---	102
11 AUG 2007	LING	7	SD	5HCP	3	12	9-6	9	H	---	98
27 AUG 2007	KEMP	7	SD	5HCP	3	14	9-3	13	P	---	94
14 SEP 2007	WOLV	6	FT	5HCP	2	13	9-3	4	P	---	99
27 SEP 2007	WOLV	6	SW	5HCP	7	11	9-4	6	H	---	87
21 OCT 2007	WOLV	5.1	SD	5HCP	1	12	9-2	5	H	---	102
06 NOV 2007	LING	6	SD	4HCP	1	7	8-9	7	P	---	93
19 NOV 2007	WOLV	5.1	SW	3HCP	2	13	8-4	1	P	---	109
11 DEC 2007	SOUT	5	VS	3HCP	6	10	9-3	9	P	---	98
28 DEC 2007	LING	6	SD	2HCP	1	9	8-1	3	H	---	107
19 JAN 2008	LING	6	SD	2HCP	2	11	8-8	2	P	+5	107

The raced panned out perfectly for Ebraam, who sat well off the furious pace early before being delivered late and getting up to win by a neck. Silver Prelude won the battle for the early lead and kept on going for longer than might have been expected to finish a close third, but Maltese Falcon responded badly to not being able to get to the front and dropped away to finish a well-beaten eighth.

Whilst describing the scenario surrounding Chia earlier, I mentioned a mare by the name of Casablanca Minx. She is only modest (officially rated 55 at the time of writing) but she does have a distinctive profile. I had long held a theory about her – I believed that she needed a double-figure field, a trip of around a mile at Wolverhampton, and a strong pace. Whilst my views about the size of the field and the course and distance were there in black and white for all to see, my view on the required pace was more theoretical. That is, until sectional times became more widely available.

This is the part of her profile covering the period for which sectional times were available. Unfortunately they weren't for the race in which she finished last of four behind Chia on January 10th, but as I said before the early pace was pedestrian and had it been possible to calculate a sectional rating it would have been a very slow one indeed. As it is, two of her best efforts in the first half of 2008 came in races with fast sectional ratings, and her only win came in the race with the quickest sectional rating of –19 (19 lengths per mile faster than the standard first sectional split).

02 JAN 2008	WOLV	8.6	SW	5HCP	6	9	8-2	10	H	+7	98
04 JAN 2008	WOLV	9.5	SW	6HCP	5	9	8-12	3	H	+12	92
10 JAN 2008	WOLV	9.5	VS	5HCP	4	4	8-1	4	H	---	65
04 FEB 2008	LING	8	SD	5HCP	4	4	8-9	2	P	+43	84
08 FEB 2008	WOLV	9.5	SW	6HCP	2	8	8-10	4	H	+50	92
15 FEB 2008	WOLV	8.6	SW	5CLM	3	6	8-4	2	P	+43	90
23 FEB 2008	WOLV	8.6	SW	6HCP	4	6	8-12	1	H	+14	97
28 FEB 2008	LING	10	SD	5HCP	8	8	8-10	7	H	+2	94
08 MAR 2008	WOLV	9.5	SW	6HCP	4	9	8-13	3	H	+6	94
13 MAR 2008	WOLV	8.6	SW	6HCP	5	11	8-6	3	H	-18	102
17 MAR 2008	WOLV	8.6	SD	6CLM	2	11	10-8	9	H	-13	93
26 MAR 2008	WOLV	9.5	SD	6HCP	1	10	8-5	8	H	-19	101
14 APR 2008	WOLV	9.5	SW	6HCP	6	7	8-7	2	H	+55	87
16 APR 2008	WOLV	8.6	VS	6HCP	3	10	8-6	8	H	+39	97
19 APR 2008	WOLV	8.6	SW	5HCP	7	8	8-6	1	P	+26	90
08 JUL 2008	WOLV	8.6	SW	6SEL	2	12	9-9	4	H	---	101

Ebraam, seen here winning at Lingfield in December 2007, was well suited by the strong pace when winning back at the same track two months later

One thing worth bearing in mind is that when you may have spotted a race where it seems that two or more trailblazers are likely to spoil each other's chances, there is also a strong possibility that the connections of at least one of the horses involved have spotted it too. As a result, it has been known for a horse's normal running style to be altered, the idea being that by avoiding a speed duel its chances will be improved. Unfortunately, the evidence is that it doesn't usually work. If a horse is effective under a positive ride then you can hardly expect it to be just as effective under contrasting tactics. All that happens is that by avoiding an early battle it just results in the other pacesetter having the stage to himself, and very possibly benefiting from it.

Unfortunately, in this situation we don't know how things are going to pan out until the race is under way, but unlike in days gone by we now have betting in running and this is another area where this form of punting can be especially productive.

PUTTING IT ALL TOGETHER

HAVING collated all the information deemed most relevant in order to start making some selections and then possibly place a bet, how do you go about using it?

Not everyone is computer literate and even though I use computers for my work and my hobby all the time, I actually hate them. Years ago I would use sheets of A4 paper, a pen and a calculator to work out my ratings and keep records, or I would write the ratings in the pullout results section of one of the weekly racing newspapers. Each to their own, but whatever you do you will need somewhere to keep all your ratings, draw statistics, front-runners' statistics, pace and track biases, sectional time data, etc. I understand that not all of the aforementioned factors will float everyone's boat, but the more of them you have, especially if they are truly original, the more of a significant edge you will gain. Obviously a computerised formbook would make life so much easier if you wish to go to some expense, or if like me you want to create some sort of database from scratch, so much the better.

This is the sort of information I can get from mine.

I create a sort of racecard containing all of the details I need about an upcoming race, and the past sand performances of the horses taking part in it. Each line of form contains many of the factors described in the previous chapters, like this -

03MAR2008	LING	5	SD	53	6HCP	2	8	8-11	2	H	-4	104

This line of form is nothing like as complicated as it may look. In fact it is very straightforward.

The first column is pretty obvious; it's the date of the race. The next column is the name of the track followed by the distance of the race in furlongs. After that is the state of the ground as I have assessed it through the calculation of the speed figures (VF – Very Fast, FT – Fast, SD – Standard, SW – Slow, VS – Very Slow). The next figure is the Pace Rating for that fixture – in this case 53 which means that there was a very slight bias towards prominent racers, but nothing too significant.

The next column is a four-character description of the type of race, in this case '6HCP' which means a Class 6 handicap. Other race types are: MDN – Maiden, CLM – Claimer, CND – Conditions Race, SEL – Seller, LST – Listed Race, GRP – Group Race, CLF – Classified Race.

The next column is the finishing position followed by the number of runners, so this horse had finished second of eight. Then follows the actual weight the horse carried, minus any allowances and plus any overweight, followed by the draw the horse started from. Next is one of the pace symbols defining the position the horse took early in the race (L – Led, P – Prominent, H – Held Up).

Next is the Sectional Rating – in this case the rating is –4 which means that the field covered the first sectional a net four lengths per mile faster than average, which is reasonably fast and suggests a truly run race.

The last figure is the speed rating.

So now that we are armed with a plethora of information, what use can it be? The following races are good examples of when this sort of information proved very useful.

Take a look at the following Class 7 event (formerly known as a banded race) from a meeting at Kempton early in 2008.

Note: In all of the example races shown in this chapter, only All-Weather form since the start of 2007 is shown in order to save on space, but originally each horse's entire record on sand was included.

6-20 KEMPTON.CO.UK CLASSIFIED STAKES (0-45) (7) 4YO+

5 furlongs Par = 99 **8 runners**

1 7 SAVILE'S DELIGHT	9-6		TOM DASCOMBE				RICHARD KINGSCOTE						
	08JAN2007	WOLV	6	SD	42	6HCP	8	10	8-6	2	L	---	85
	05FEB2007	WOLV	5.1	SD	25	6HCP	8	11	8-3	11	P	---	92
	13FEB2007	SOUT	6	SD	55	6HCP	9	14	8-12	7	P	---	89
	22FEB2007	SOUT	7	FT	48	6HCP	14	14	8-12	1	P	---	71
	20MAR2007	SOUT	7	SW	62	7CLF	4	12	9-0	3	P	---	94
	08MAY2007	SOUT	7	SW	53	6HCP	7	12	8-8	2	H	---	86
	11DEC2007	SOUT	6	SD	39	6HCP	6	10	8-5	9	P	---	90
	14JAN2008	SOUT	5	VS	50	6HCP	1	9	11-2	9	H	+6	101
2 1 BLUSHING RUSSIAN	9-0(B)		J M BRADLEY				TOLLEY DEAN(5)						
	03JAN2007	KEMP	8	SW	50	7CLF	8	13	9-0	7	P	---	86
	15JAN2007	WOLV	5.1	SD	55	6HCP	5	10	10-9	10	H	---	84
	28JAN2007	WOLV	5.1	SD	50	6HCP	1	9	8-8	7	L	---	99
	25FEB2007	KEMP	6	SD	37	6HCP	10	11	8-10	5	P	---	92
	28FEB2007	SOUT	6	SD	58	6HCP	6	11	8-8	7	P	---	87
	16MAR2007	LING	6	SD	55	6HCP	6	11	8-8	11	P	---	91
	23MAR2007	WOLV	5.1	SD	45	6HCP	11	12	8-5	3	H	---	89
	13APR2007	SOUT	6	SW	58	6SEL	8	12	9-6	6	P	---	93
	16APR2007	WOLV	5.1	SD	42	6SEL	6	13	9-3	13	H	---	90
	03DEC2007	WOLV	5.1	SW	38	6HCP	8	10	8-1	3	H	---	89
	19DEC2007	LING	6	SD	48	6HCP	5	12	7-13	8	H	---	96
	02JAN2008	WOLV	6	SW	52	6SEL	8	12	8-9	9	H	-20	89
	05JAN2008	KEMP	5	SD	55	7CLF	4	11	9-0	10	P	---	96
	09JAN2008	WOLV	5.1	VS	42	7CLF	2	12	8-9	6	P	-9	94
	14JAN2008	SOUT	5	VS	50	6HCP	3	9	10-12	3	L	+6	95
3 3 CHARLOTTE GREY *Non-Runner*													
4 8 DODAA	9-0		N WILSON				ASHLEY HAMBLETT(5)						
	25JAN2007	SOUT	6	FT	72	7CLF	3	11	9-0	2	L	---	94
	01FEB2007	SOUT	6	SW	62	7CLF	3	14	9-0	4	L	---	94
	22FEB2007	SOUT	5	SW	--	7CLF	1	14	9-0	2	P	---	101
	15MAR2007	SOUT	5	VS	--	6HCP	8	13	8-9	10	H	---	97

	22OCT2007	WOLV	6	SD	36	7CLF	7	13	8-13	3	P	---	85
	30OCT2007	WOLV	5.1	SD	62	7CLF	7	13	9-0	12	H	---	83
	11DEC2007	SOUT	6	SD	39	6HCP	10	10	8-5	8	P	---	75
	14DEC2007	SOUT	5	VS	--	6HCP	5	13	8-6	8	L	---	89
	31DEC2007	WOLV	5.1	SW	62	6HCP	5	10	8-7	7	P	---	93
	09JAN2008	WOLV	5.1	VS	42	7CLF	1	12	9-0	3	L	-9	97
5 2 ELIZABETH SPIRIT	9-0		E S MCMAHON				STEPHEN DONOHOE						
	20OCT2007	WOLV	5.1	SD	47	5MDN	9	12	8-9	8	H	---	83
	28NOV2007	KEMP	6	SD	48	6MDN	6	12	8-12	12	L	---	90
	13DEC2007	WOLV	6	VS	44	6MDN	5	8	8-12	2	L	---	88
6 6 ELVINA	9-0		A G NEWCOMBE				FERGUS SWEENEY						
	21JAN2007	WOLV	5.1	SD	48	7CLF	1	13	9-0	8	P	---	95
	15MAR2007	SOUT	5	VS	--	6HCP	4	13	8-10	5	P	---	100
	21MAY2007	WOLV	5.1	SD	33	6HCP	10	12	8-7	9	H	---	90
	30AUG2007	WOLV	5.1	SD	67	6HCP	8	12	8-9	3	P	---	96
	20SEP2007	WOLV	5.1	SW	50	6HCP	5	11	9-0	4	P	---	95
	05JAN2008	KEMP	5	SD	55	7CLF	5	11	9-0	2	L	---	95
7 5 LADY HOPEFUL	9-0(B)		PETER GRAYSON				L P KENIRY						
	15JAN2007	WOLV	5.1	SD	55	6HCP	4	10	10-9	3	H	---	84
	21JAN2007	WOLV	5.1	SD	48	7CLF	3	13	9-0	3	H	---	92
	28JAN2007	WOLV	5.1	SD	50	6HCP	2	10	8-8	7	H	---	101
	12FEB2007	WOLV	5.1	SD	40	6HCP	8	11	8-11	4	H	---	92
	05MAR2007	LING	5	SD	38	6HCP	5	10	8-7	6	H	---	98
	16MAR2007	LING	6	SD	55	6HCP	5	11	8-9	5	H	---	94
	23MAR2007	WOLV	5.1	SD	45	6HCP	5	12	8-6	13	H	---	95
	01MAY2007	WOLV	5.1	FT	52	6HCP	5	11	8-9	12	H	---	94
	31MAY2007	LING	5	SD	36	7CLF	4	10	9-0	5	H	---	92
	02JUL2007	WOLV	5.1	SD	33	6SEL	4	12	8-8	2	H	---	98
	30AUG2007	WOLV	5.1	SD	67	5SEL	8	13	8-7	8	P	---	87
	20SEP2007	WOLV	6	SW	50	7CLF	9	13	9-0	5	H	---	89
	19NOV2007	KEMP	5	FT	50	7CLF	5	12	9-0	3	H	---	95
	03DEC2007	WOLV	5.1	SW	38	6HCP	7	10	8-4	2	P	---	91
	31DEC2007	WOLV	5.1	SW	62	6HCP	4	7	8-7	9	H	---	95
	05JAN2008	KEMP	5	SD	55	7CLF	3	11	9-0	3	H	---	96
8 4 PRETTILINI	*Non-Runner*												

Draw and Front Runner Statistics for 5 Furlongs during past 12 months

Draw	Pos	Winners %	
Low	7	23	
Middle	9	29	
High	15	48	
FROM RAIL	**RUNNERS**	**WINNERS**	**%**
1	29	8	28
2	31	3	10
3	28	3	11
4	29	6	21
5	29	0	0
6	26	0	0
7	27	3	11
8	23	2	9

Front Runners Impact Value = 3.2
Prominent Horses Impact Value = 0.9
Hold Up Horses Impact Value = 0.6

As you can see, quite a few of the runners involved in this race had led early at least once in their recent races, but one of them – Dodaa – had done it rather more often. He had won a similar race to this recently, which had been truly run (Sectional Rating –9) and had done so on a track that had a slight bias against front-runners (42).

The key to his prospects, however, was that plum draw in stall eight of eight. Just look at those inverted draw statistics – the rails draw was currently boasting a record of just under one winner in three starting from the stall, whilst the Impact Values show that early leaders were winning more than three times more often as they were entitled to. He just had to be backed and those who did never had a moment's worry as he made every yard to win unchallenged at odds of 2-1.

Dodaa will make another appearance later on in this book and the scenario will not be unlike the one described for this race.

The next contest is another five-furlong event at Kempton, which took place the following month, and if ever there was a race that demonstrated the value of doing your homework and having statistics close at hand, then this was it.

1-50 PANORAMIC BAR & RESTAURANT HANDICAP (0-60) (6) 4YO+

5 furlongs Par = 101 12 runners

1 9 MAMBAZO 9-4(E/S) S C WILLIAMS WILLIAM CARSON(7)

08JAN2007	WOLV	6	SD	42	5HCP	5	8	8-13	7	H	---	93
10JAN2007	LING	6	SD	43	5HCP	6	11	8-8	10	H	---	101
27JAN2007	WOLV	5.1	SW	22	5HCP	9	9	8-6	6	H	---	83
16FEB2007	WOLV	6	SD	52	4HCP	7	10	7-11	10	H	---	98
26FEB2007	WOLV	5.1	SD	57	5HCP	6	12	7-11	8	H	---	100
21APR2007	WOLV	6	SD	58	5HCP	10	11	8-7	9	P	---	95
23MAY2007	LING	6	SD	40	6HCP	3	12	9-4	4	H	---	98
11JUN2007	WOLV	5.1	SD	44	6HCP	3	13	9-2	5	H	---	99
21JUN2007	LING	6	SW	44	6HCP	1	11	9-3	9	P	---	99
29JUN2007	WOLV	6	SD	61	6HCP	1	12	9-7	4	P	---	101
03JUL2007	LING	6	SD	36	4HCP	3	8	8-12	9	P	---	92
04JUL2007	KEMP	6	SD	64	5HCP	1	8	9-6	6	L	---	99
10JUL2007	LING	6	SD	47	5HCP	7	11	9-9	9	H	---	98
30AUG2007	LING	6	SD	63	4HCP	9	12	8-12	5	L	---	97
31OCT2007	LING	5	SD	48	4HCP	9	10	8-4	4	H	---	98
16NOV2007	LING	5	SD	52	5HCP	9	9	8-8	10	H	---	92
22DEC2007	LING	5	SD	64	5HCP	6	9	9-2	7	H	---	87
13JAN2008	KEMP	6	SD	43	5HCP	9	10	8-13	8	H	---	92
01FEB2008	KEMP	6	SW	48	6HCP	2	8	8-11	8	L	---	101

2 10 STONEACRE BOY 9-4 PETER GRAYSON L P KENIRY

06JAN2007	KEMP	5	SD	45	5HCP	5	9	9-2	6	P	---	96
17JAN2007	LING	5	SD	57	5HCP	9	9	8-3	7	H	---	90
11FEB2007	SOUT	5	SD	--	5HCP	2	10	8-10	7	H	---	103
06MAR2007	SOUT	5	SD	--	5HCP	2	11	9-1	6	P	---	102
22MAR2007	SOUT	5	SW	--	5HCP	8	10	8-10	2	P	---	83
03APR2007	SOUT	5	SW	--	5HCP	7	9	9-1	5	H	---	94
08MAY2007	SOUT	5	FT	--	5HCP	11	11	8-8	6	H	---	55
10JUL2007	LING	6	SD	47	5HCP	10	11	9-2	10	H	---	83
20SEP2007	WOLV	5.1	SW	50	6HCP	6	12	8-13	3	H	---	97
13OCT2007	KEMP	5	FT	33	6HCP	1	12	8-10	9	P	---	103
17NOV2007	WOLV	5.1	SW	56	5HCP	6	9	8-11	6	P	---	91
12DEC2007	KEMP	5	SD	62	5HCP	6	6	8-5	4	H	---	78
03JAN2008	WOLV	5.1	SW	47	6HCP	10	10	9-1	5	P	-11	76

3 1 CAMISSA	9-4		D K IVORY				JAMES O'REILLY(5)						
	23MAY2007	LING	5	SD	40	4HCP	7	9	8-13	2	H	---	90
	10JUL2007	LING	8	SD	47	5CLM	5	11	8-12	11	H	---	91
4 5 PERLACHY *Non-Runner*													
5 8 RADIATOR ROONEY	9-2(V)		PATRICK MORRIS				JAMIE SPENCER						
	06JAN2007	KEMP	5	SD	45	5HCP	8	9	8-10	4	P	---	91
	05FEB2007	LING	6	SD	45	5HCP	3	11	7-13	8	P	---	104
	15FEB2007	SOUT	6	SD	50	6HCP	8	10	9-2	3	P	---	82
	20OCT2007	WOLV	8.6	SD	47	6HCP	7	13	9-3	5	P	---	94
	04DEC2007	WOLV	7.1	SW	38	6HCP	5	11	8-5	10	P	---	93
	20DEC2007	WOLV	5.1	VS	42	6HCP	4	13	8-4	8	H	---	102
	05JAN2008	KEMP	6	SD	55	6HCP	3	10	9-2	12	P	---	98
	02FEB2008	KEMP	6	SD	38	5HCP	1	8	8-6	6	P	+24	105
6 6 KEMPSEY	9-2(B)		J J BRIDGER				CHRIS CATLIN						
	06JAN2007	KEMP	5	SD	45	5HCP	9	9	7-13	10	H	---	90
	10JAN2007	KEMP	5	SW	38	6HCP	1	12	8-9	12	P	---	99
	17JAN2007	LING	6	SD	57	6HCP	3	10	8-11	7	P	---	101
	28JAN2007	KEMP	6	SD	40	5HCP	4	11	8-5	4	L	---	99
	25FEB2007	KEMP	6	SD	37	6HCP	4	11	8-7	9	L	---	105
	06MAR2007	SOUT	5	SD	--	5HCP	4	11	8-7	3	P	---	99
	26MAR2007	KEMP	5	SD	52	5HCP	3	9	8-7	2	L	---	99
	13JAN2008	KEMP	7	SD	43	6HCP	8	10	8-12	6	H	---	93
	26JAN2008	LING	6	SD	50	5HCP	3	5	8-4	4	L	0	101
7 11 OVERSTAYED	9-1(EB)		M MULLINEAUX				JOHN MCAULEY(3)						
	08DEC2007	WOLV	5.1	SW	50	4HCP	7	8	8-10	8	P	---	95
	20DEC2007	WOLV	5.1	VS	42	6HCP	7	13	9-4	7	P	---	101
	25JAN2008	WOLV	5.1	SD	48	6HCP	5	6	8-11	6	P	+3	96
8 4 REGAL ROYALE	8-13(V)		PETER GRAYSON				ADRIAN MCCARTHY						
	27JAN2007	LING	8	FT	52	4HCP	7	9	9-1	5	H	---	94
	10FEB2007	LING	6	SD	48	4HCP	6	12	8-7	5	H	---	106
	27FEB2007	LING	8	SD	43	3HCP	8	8	8-5	3	P	---	98
	17MAR2007	LING	7	SD	38	4HCP	13	13	8-12	6	H	---	93
	24MAR2007	KEMP	7	SD	52	4HCP	6	9	8-6	2	P	---	101
	21APR2007	WOLV	6	SD	58	5HCP	9	11	9-4	10	P	---	96

	25APR2007	KEMP	6	SD	55	5HCP	11	11	9-0	10	P	---	94
	10JUL2007	LING	6	SD	47	5HCP	4	11	9-5	11	H	---	100
	25JUL2007	LING	6	SD	60	5HCP	10	11	9-7	7	P	---	94
	09NOV2007	WOLV	7.1	SD	47	6HCP	12	12	9-2	9	H	---	81
	24NOV2007	LING	6	FT	65	6HCP	9	11	9-2	9	H	---	96
	12DEC2007	KEMP	6	SD	62	6CLF	3	8	9-0	5	P	---	99
	19DEC2007	LING	6	SD	48	6HCP	3	12	8-13	1	P	---	98
	29DEC2007	LING	6	SD	33	6HCP	10	11	9-0	7	H	---	89
	30DEC2007	LING	7	SD	36	6HCP	7	11	9-0	7	H	---	87
	09JAN2008	KEMP	6	SD	64	6CLF	2	7	8-9	7	P	---	98
	11JAN2008	KEMP	6	FT	57	6HCP	2	8	8-11	6	L	-1	102
	18JAN2008	KEMP	6	SD	57	6HCP	1	9	8-9	8	L	+3	100
9 7 RACING STRIPES	8-10(B)			K O CUNNINGHAM-BROWN					NEIL CHALMERS(3)				
	23OCT2007	LING	6	SD	50	5HCP	11	11	9-5	1	P	---	74
	01DEC2007	KEMP	8	SD	48	5HCP	11	13	8-9	12	H	---	82
	16DEC2007	KEMP	7	FT	21	5HCP	14	14	8-5	11	L	---	80
	18JAN2008	KEMP	6	SD	57	6HCP	6	9	8-10	5	H	+3	95
	23JAN2008	KEMP	5	SD	57	6HCP	5	7	8-10	3	H	+8	98
10 3 DODAA *Non-Runner*													
11 12 SIR LOIN	8-8(V)			P BURGOYNE				ROSS ATKINSON(7)					
	10JAN2007	KEMP	5	SW	38	6HCP	4	12	8-8	8	H	---	97
	26JAN2007	WOLV	5.1	SW	68	5CLM	6	12	9-11	10	P	---	88
	01FEB2007	SOUT	5	SD	--	6HCP	8	11	8-5	9	P	---	84
	26FEB2007	WOLV	5.1	SD	57	5HCP	11	12	7-9	7	P	---	93
	01JUN2007	WOLV	5.1	SD	67	6HCP	1	12	8-8	9	P	---	102
	11JUN2007	WOLV	5.1	SD	44	6HCP	12	13	8-12	8	P	---	88
	02JUL2007	WOLV	5.1	SD	33	6SEL	8	12	9-2	11	H	---	90
	04OCT2007	KEMP	7	FT	53	6HCP	9	13	9-0	7	H	---	83
	17OCT2007	KEMP	6	SD	36	6HCP	11	11	8-12	6	L	---	73
	29OCT2007	KEMP	5	SD	57	6HCP	3	10	8-9	3	P	---	101
	11NOV2007	KEMP	5	FT	52	6HCP	2	12	9-0	11	L	---	103
	27NOV2007	WOLV	5.1	SW	38	6HCP	7	13	8-13	5	L	---	93
	09DEC2007	LING	5	SD	50	6HCP	10	10	8-11	9	H	---	85
	19DEC2007	LING	6	SD	48	6HCP	4	12	8-5	2	L	---	97
	29DEC2007	LING	6	SD	33	6HCP	4	11	8-11	5	L	---	98
	09JAN2008	KEMP	5	SD	64	6HCP	1	9	7-12	3	L	---	99
	23JAN2008	KEMP	5	SD	57	6HCP	4	7	8-2		L	+8	98

12 2 KING OF CHARM 8-7(B) G L MOORE FERGUS SWEENEY

02FEB2007	WOLV	6	SD	40	6HCP	8	13	8-10	4	H	---	88
07FEB2007	LING	7	FT	50	6HCP	6	9	8-8	10	H	---	95
25FEB2007	KEMP	6	SD	37	6HCP	5	11	8-9	6	P	---	97
16MAR2007	LING	6	SD	55	6HCP	4	11	8-7	6	P	---	95
21MAR2007	KEMP	6	SD	65	6HCP	1	12	8-11	11	P	---	95
20APR2007	LING	6	SD	47	6HCP	10	12	8-11	8	H	---	90
02MAY2007	KEMP	6	SD	55	6HCP	6	12	9-4	7	P	---	92
22OCT2007	WOLV	6	SD	36	7CLF	12	13	8-13	10	H	---	74
29OCT2007	LING	6	SD	19	6HCP	6	12	8-0	6	P	---	92
26NOV2007	LING	7	SD	46	6HCP	7	11	7-12	3	P	---	93
04DEC2007	LING	6	SD	36	6HCP	6	12	8-5	5	H	---	94
11JAN2008	KEMP	6	FT	57	7CLF	8	8	8-9	2	P	+18	73
16JAN2008	KEMP	6	SD	40	7CLF	3	9	9-0	6	P	+5	93
25JAN2008	KEMP	5	SD	62	7CLF	1	6	9-0	6	P	+20	94

Draw and Front Runner Statistics for 5 Furlongs during past 12 months

Draw Pos	Winners	%	
Low	8	21	
Middle	12	31	
High	19	49	
FROM RAIL	**RUNNERS**	**WINNERS**	**%**
1	36	12	33
2	39	4	10
3	35	5	14
4	35	6	17
5	35	0	0
6	30	0	0
7	30	3	10
8	24	3	13
9	23	4	17
10	17	1	6
11	15	0	0
12	10	1	10

Front Runners Impact Value = 3.2
Prominent Horses Impact Value = 0.8
Hold Up Horses Impact Value = 0.6

When you see a race in this format, it quickly becomes obvious that several of these possessed plenty of early pace, but the daddy of them all was Sir Loin. The seven-year-old gelding was just about the fastest horse around over four furlongs and if races took place over that distance in this country then he would have a very impressive win record. Even over five furlongs, he could win races provided things went his way even though the last furlong could be a bit of a problem more often than not.

He had been successful in his next-to-last start over this course and distance and the comment-in-running just about said it all.

made all: ridden over 1f out: coaxed along final furlong: just held on

So how would he cope today? It's true that he was up against one or two others that had been a bit smart from the gate and could be a problem to him if they got him at it early on. However, as with Dodaa in the previous example, he had the plum rails draw and that would make things a lot easier. The five-furlong start at Kempton is quite close to the first bend and I could visualise him railing like a greyhound and establishing enough of an advantage to see him home up the home straight (the five-furlong course at Kempton uses the inner home bend with a shorter run-in). It all looked so straightforward, which makes the way the betting market unfolded especially fascinating.

The Irish-trained Radiator Rooney had won at the same track the previous day and this time had the services of Jamie Spencer, but that win was over six furlongs and he was up against some real trailblazers here. Wasn't there a real possibility that he could get himself outpaced? Also, although the winning time of the contest he won the day before was sound enough (speed rating 105) the early pace had been very slow (Sectional Rating +24) which was another worry if today's race developed into an early burn-up. Add to that a record of just two wins from 39 Flat outings then surely odds of 5-4 about Radiator Rooney were far too skinny.

Even more interesting was Mambazo. I was there when he ran at the same track just two days earlier and had backed him to win a six-furlong handicap. On that occasion he had started from the rails draw (also a big advantage over that trip) and after breaking fast from the stalls he went off at a million miles an hour. He managed to maintain the advantage for a long way and although he was eventually reeled in it was only by one horse, the favourite Littledodayno.

It appeared to be so simple didn't it? After all he had blitzed his way around the same track two days earlier and nearly got home over six furlongs, so over five furlongs he would probably see it out with the same tactics applied, wouldn't he?

Well there were a couple of problems. Firstly, although he was quite well drawn here, he wasn't *the* best drawn horse unlike last time. Also being able to dominate seven rivals over six furlongs was one thing, but this time he was up against horses with even greater early speed than him and the most potent of them all had the plum draw. Here was a situation where people had become blinkered by what had happened at the same track two days before and not looked beyond that. Hence the reason he went off the well-backed 7-2 second favourite.

The way things eventually panned out could have been predicted by anyone that had looked hard enough to unravel this puzzle. Again the comment-in running for Sir Loin summarises the situation perfectly:

made all and soon at least two lengths clear: tired final furlong: always just holding on

To be fair to Radiator Rooney he was always in a good position behind the leaders and tried very hard, but again over this sharp five furlongs the front-runners weren't coming back and fourth place was the best he could do, but what about Mambazo? Perhaps connections had looked closely at the race themselves and worked out that they weren't going to be able to get the early lead as they had done on Friday, and if they tried to do so all they would do was cut their own throats.

Either way, jockey William Carson held him up, but he then met trouble in running and over this course and distance that was always going to prove fatal. Perhaps he should have been a bit more aggressive rather than let him coast home ninth of the ten runners – the Stewards obviously thought so as they slapped him with a 14-day ban for not taking all reasonable and permissible measures to achieve the best-possible placing – but to my mind he was a touch unlucky and perhaps a little naïve rather than anything more sinister.

The biggest surprise of the whole race, however, was how Sir Loin was allowed to go off at 8-1.

Here is another race from Southwell in June 2008 in which one of the contenders appears to hold an outstanding chance:

2-40 SHANE W DARBY MEMORIAL SELLING HANDICAP (0-60) (6) 4YO+

7 furlongs **Par=101** **14 runners**

1 9 ELUSIVE WARRIOR	9-4(P)		A J MCCABE				JAMES DOYLE						
	18OCT2007	WOLV	7.1	SD	58	6HCP	4	11	9-2	10	P	---	89
	01NOV2007	WOLV	7.1	SD	72	5HCP	2	9	8-5	7	P	---	93
	04NOV2007	WOLV	7.1	SD	55	5HCP	2	10	8-5	1	L	---	98
	04DEC2007	WOLV	7.1	SW	38	6HCP	12	12	8-11	7	P	---	91
	11DEC2007	SOUT	6	SD	39	6HCP	5	10	8-13	1	H	---	90
	07JAN2008	SOUT	7	SD	57	6HCP	3	6	8-11	6	P	+11	97
	14JAN2008	SOUT	7	SD	64	6SEL	2	5	8-12	2	P	-6	94
	11MAR2008	SOUT	5	SD	83	6HCP	7	10	8-11	5	H	-3	84
	13MAR2008	SOUT	7	SW	50	6HCP	6	6	8-6	1	P	-4	82
	26MAR2008	SOUT	7	SD	54	6HCP	1	11	8-9	9	P	-18	101
	08APR2008	SOUT	7	VS	50	5HCP	1	8	8-4	1	L	-13	105
	19APR2008	WOLV	7.1	SW	56	6HCP	7	11	9-0	2	L	-23	95
	24APR2008	SOUT	8	SD	69	5HCP	2	6	8-6	6	L	-8	92
	13MAY2008	SOUT	6	SW	63	5HCP	6	10	8-13	2	P	-25	95
	20MAY2008	SOUT	8	SD	37	5HCP	4	8	8-4	6	P	-20	100
	03JUN2008	SOUT	7	FT	57	6CLM	2	14	9-8	10	P	-23	97
2 6 UNCLE BULGARIA	8-13		G C BRAVERY				SEB SANDERS						
3 11 GOODBYE CASH	8-12		P D EVANS				TOM EAVES						
	07MAY2007	KEMP	5	SW	54	4HCP	1	7	8-0	7	H	---	95
	24SEP2007	KEMP	6	SD	54	4HCP	4	10	9-0	2	H	---	95
	06OCT2007	KEMP	7	FT	26	4HCP	10	13	8-12	5	H	---	95
	30DEC2007	LING	6	SD	36	6CLM	4	12	8-8	7	H	---	103
	06JAN2008	SOUT	6	SD	48	6CLM	2	9	8-9	5	P	---	97
	18JAN2008	KEMP	6	SD	57	4HCP	7	8	8-4	9	P	-0	97
	29JAN2008	SOUT	6	FT	60	6CLM	4	5	8-11	4	P	+25	83
	13FEB2008	KEMP	6	SW	41	5HCP	7	11	9-3	6	P	+3	101
	19FEB2008	SOUT	7	VS	50	4HCP	5	5	8-8	5	L	+11	87
	03MAR2008	WOLV	6	SW	48	6CLM	3	8	8-1	7	H	+5	96
	08MAR2008	WOLV	6	SW	53	5CLM	5	6	8-8	5	H	+43	83
	05APR2008	KEMP	6	SW	33	6CLM	3	4	9-1	1	P	+22	96

4 2 CABOURG	8-11		R BASTIMAN				ROBERT WINSTON						
	13JAN2007	WOLV	8.6	SW	42	6HCP	2	13	8-9	13	H	---	101
	02FEB2007	WOLV	8.6	SD	40	6HCP	7	9	8-9	9	H	---	86
	16FEB2007	WOLV	7.1	SD	52	6HCP	2	12	8-11	6	H	---	95
	24FEB2007	LING	7	FT	48	6HCP	4	12	9-2	9	H	---	95
	30MAR2007	WOLV	7.1	SD	50	6HCP	5	8	8-7	3	H	---	97
	21NOV2007	KEMP	7	FT	57	6HCP	4	13	9-2	1	H	---	95
	04DEC2007	WOLV	7.1	SW	38	6HCP	1	12	8-9	4	P	---	101
	29DEC2007	LING	8	SD	33	6HCP	6	10	9-5	11	P	---	93
	15FEB2008	WOLV	7.1	SW	58	6HCP	7	12	8-13	10	H	-19	98
	29FEB2008	KEMP	7	SD	63	6HCP	9	14	9-3	2	H	-13	93
	12MAR2008	WOLV	7.1	SW	56	6HCP	4	12	9-1	12	P	-15	99
	31MAR2008	SOUT	8	VS	50	6HCP	4	14	11-0	10	P	-9	92
	14APR2008	WOLV	7.1	SW	47	6HCP	11	11	8-13	10	H	-13	78
5 10 MURDOCH	8-10		E S MCMAHON				RICHARD MULLEN						
	03MAY2007	WOLV	8.6	FT	47	6MDN	13	13	8-13	4	P	---	79
	08APR2008	SOUT	8	VS	50	6HCP	9	9	9-3	5	P	-2	70
6 8 UGENIUS	8-8		P A BLOCKLEY				STEPHEN DONOHOE						
	15JAN2007	WOLV	6	SD	55	6HCP	5	10	8-0	2	P	---	87
	13MAR2007	SOUT	6	SW	52	5MDN	6	10	8-8	6	P	---	92
	24NOV2007	WOLV	8.6	SW	47	6SEL	6	9	8-7	1	P	---	80
	05DEC2007	KEMP	8	SD	67	7CLF	12	13	9-0	12	P	---	88
	04MAR2008	SOUT	7	VS	53	6HCP	1	13	8-2	8	P	-17	97
	11MAR2008	SOUT	8	VS	42	6CLM	8	14	8-7	2	P	-20	86
	13MAR2008	SOUT	8	SW	50	6SEL	7	8	10-0	7	L	+19	81
	03JUN2008	SOUT	7	FT	57	6CLM	3	14	9-2	4	H	-23	93
7 3 COMPULSION	8-8		PAT EDDERY				PAUL HANAGAN						
	31MAY2007	LING	6	SD	36	5MDN	5	10	9-4	5	H	---	96
	05SEP2007	KEMP	7	FT	55	6MDN	6	11	9-2	9	P	---	90
	31OCT2007	KEMP	6	FT	33	6MDN	3	5	8-13	6	L	---	96
	28NOV2007	KEMP	6	SD	48	6MDN	7	12	8-12	1	H	---	86
	16FEB2008	KEMP	6	SD	45	6HCP	3	8	9-0	2	H	+7	102
	29FEB2008	KEMP	7	SD	63	6HCP	5	14	8-12	11	P	-13	99
	19MAR2008	LING	7	SD	36	6HCP	10	10	8-13	5	L	+6	84
	02APR2008	LING	6	SD	50	6SEL	6	8	9-3	9	H	+12	95

8 14 JELLYTOT	8-7		J O'REILLY				CATHERINE GANNON						
	16JAN2007	SOUT	7	VF	55	5HCP	5	8	8-7	2	H	---	96
	15MAR2007	SOUT	7	SW	43	6HCP	13	14	9-0	2	H	---	69
	01MAY2007	SOUT	7	SW	56	6HCP	6	12	8-13	4	P	---	86
9 13 DOMESDAY	8-4(P)		W G HARRISON				DURAN FENTIMAN(3)						
	27SEP2007	WOLV	9.5	SW	44	6HCP	4	13	8-12	5	L	---	97
	02NOV2007	WOLV	8.6	SD	72	6HCP	8	13	8-6	10	H	---	94
10 12 DANCE IN STYLE	8-4		A CROOK				ROYSTON FFRENCH						
11 1 HI SPEC	8-4(P)		MISS M E ROWLAND				GREG FAIRLEY						
	28SEP2007	DUND	8	SD	40	6HCP	B	13	8-10	12	H	---	0
	05DEC2007	DUND	7	SW	40	6HCP	9	14	8-10	1	H	---	89
	20JAN2008	WOLV	7.1	SW	57	7CLF	3	10	9-0	2	H	---	93
	31JAN2008	SOUT	8	SD	64	7CLF	9	12	8-9	4	P	+9	68
	23FEB2008	WOLV	7.1	SW	67	7CLF	1	11	9-0	5	P	-6	94
	28MAR2008	WOLV	8.6	SD	61	6HCP	4	12	8-3	1	P	-15	98
	11APR2008	WOLV	7.1	SW	36	6HCP	10	10	8-6	11	P	+4	78
	29APR2008	SOUT	8	SW	47	6CLM	5	7	8-4	1	P	+2	87
	06MAY2008	SOUT	6	SD	67	6HCP	6	13	8-8	4	H	---	87
12 5 CRAFTY FOX	8-4(V)		JOHN A HARRIS				DEAN MCKEOWN						
	15JAN2007	WOLV	7.1	SD	55	6SEL	3	11	9-3	3	P	---	95
	19FEB2007	LING	8	SD	25	6SEL	9	11	10-0	9	L	---	90
	25FEB2007	KEMP	7	SD	37	5HCP	7	10	7-9	10	P	---	99
	16MAR2007	LING	6	SD	55	6HCP	11	11	8-10	9	H	---	80
	01APR2007	LING	7	SD	40	6HCP	8	13	7-12	6	P	---	94
	14MAY2007	SOUT	6	FT	55	7CLF	1	12	9-0	4	H	---	98
	05JUN2007	LING	7	SW	---	6CLM	4	14	9-10	8	P	---	102
	03JUL2007	LING	8	SD	36	6HCP	8	12	9-5	3	P	---	93
	17JUL2007	KEMP	7	FT	64	5CLM	3	10	9-3	12	L	---	86
	12SEP2007	KEMP	8	SD	43	6CLM	11	11	9-4	13	L	---	80
	31MAR2008	SOUT	8	VS	50	6HCP	10	14	10-8	11	H	-9	80
13 4 UNION JACK JACKSON	8-4(B)		JOHN A HARRIS				DALE GIBSON						
	09JAN2007	SOUT	6	SD	68	6HCP	12	14	8-10	12	H	---	88

	25JAN2007	SOUT	6	FT	72	6HCP	5	10	8-12	5	P	---	92
	06MAR2007	SOUT	6	FT	57	6HCP	9	14	8-7	11	H	---	86
	27MAR2007	WOLV	7.1	FT	45	6HCP	1	12	8-9	4	H	---	95
	20APR2007	LING	6	SD	47	6HCP	1	12	8-9	2	P	---	100
	28APR2007	WOLV	6	SD	63	6HCP	2	11	10-9	1	H	---	97
	15MAY2007	SOUT	6	FT	58	6HCP	1	14	9-3	5	P	---	100
	23MAY2007	LING	7	SD	40	6HCP	5	14	9-5	3	H	---	94
	05JUN2007	SOUT	7	FT	47	4HCP	4	5	8-2	1	L	---	92
	10JUL2007	WOLV	6	SD	50	6HCP	9	12	9-6	5	H	---	93
	04FEB2008	SOUT	5	SW	38	6HCP	8	8	9-4	6	P	+2	32
	17FEB2008	SOUT	5	VS	67	6HCP	11	12	8-6	13	H	---	66
	04MAR2008	SOUT	7	VS	53	6HCP	13	13	8-12	11	H	-17	60
	13MAR2008	WOLV	8.6	SW	58	6HCP	11	11	8-4	6	H	-18	81
	28APR2008	SOUT	6	SW	70	6HCP	7	14	8-7	6	H	-13	91
	06MAY2008	SOUT	6	SD	67	6HCP	3	13	8-8	14	H	---	90
14 7 PAJADA	8-4(EB)		M D I USHER				JUSTINA KAY(7)						
	04FEB2007	KEMP	7	FT	32	6HCP	14	14	8-3	1	H	---	76
	14FEB2007	LING	10	FT	38	5HCP	7	9	8-4	3	P	---	84
	27FEB2007	LING	7	SD	43	6HCP	9	14	8-6	8	P	---	82
	20APR2007	LING	8	SD	47	7CLF	4	12	8-7	7	P	---	92
	01MAY2007	WOLV	8.6	FT	52	7CLF	4	13	8-7	12	P	---	92
	14OCT2007	LING	10	SW	33	6HCP	8	12	8-2	12	P	---	84
	11NOV2007	KEMP	8	FT	52	6MDN	10	14	8-12	3	P	---	88
	16NOV2007	LING	7	SD	52	6HCP	6	14	7-11	1	P	---	97
	26NOV2007	LING	7	SD	46	6HCP	10	11	8-4	2	L	---	89
	09DEC2007	LING	10	SD	50	6HCP	10	11	8-4	3	H	---	87
	10FEB2008	KEMP	8	SD	52	7CLF	10	10	9-0	7	P	---	89
	20FEB2008	KEMP	7	SD	55	7CLF	8	10	8-7	4	H	+9	93
	27FEB2008	KEMP	8	SD	64	6HCP	7	9	8-3	4	H	+26	86
	28MAR2008	KEMP	7	SD	45	7CLF	7	13	8-7	8	P	+6	89
	09APR2008	KEMP	8	SD	50	7CLF	6	14	8-7	8	P	-4	90
	29APR2008	SOUT	7	SW	47	6HCP	4	9	8-2	9	P	-14	93

Draw and Front Runner Statistics for 7 Furlongs during past 12 months

Draw Pos	Winners	%
Low	17	31
Middle	20	37
High	17	31

DRAW NO	RUNNERS	WINNERS	%
1	51	4	8
2	48	6	13
3	52	11	21
4	50	8	16
5	49	3	6
6	44	11	25
7	37	4	11
8	27	4	15
9	22	3	14
10	17	0	0
11	11	0	0
12	6	0	0
13	4	0	0
14	3	0	0

Front Runners Impact Value = 1.0
Prominent Horses Impact Value = 1.3
Hold Up Horses Impact Value = 0.6

This race was rather more straightforward and less controversial than the previous contest. Basically Elusive Warrior had so much going for him that to use a cliché he had ticks in all the right boxes.

Look at his profile. The course and distance were absolutely perfect, he had earned good speed figures in a series of fast-run races at this track, his draw was fine and his style of running was ok too. If only every race was as predictable as this we would all have few worries. At times like this you just have to open the wallet and bet without fear. Those that did were rewarded with an easy 9-4 winner.

Below is a race run at Lingfield in June 2008 where the previous speed figures earned by those taking part helped narrow down a tricky looking contest to just a couple of possible winners.

6-25 MARSH GREEN HANDICAP (0-60) (6) 4YO+

10 furlongs **Par = 101** **14 runners**

1 5 KINGS TOPIC	9-4(P)		A B HAYNES			SEB SANDERS							
	13AUG2007	WOLV	9.5	SD	42	6HCP	10	11	8-13	2	P	---	90
	18AUG2007	LING	10	SD	---	6HCP	3	12	9-9	7	H	---	100
	08SEP2007	WOLV	8.6	SD	44	6HCP	13	13	9-10	5	H	---	80
	19NOV2007	KEMP	10	FT	50	6HCP	10	10	9-8	6	H	---	52
	27MAR2008	LING	10	SD	44	5HCP	3	8	8-0	1	H	+15	102
	02APR2008	LING	10	SD	50	6HCP	2	6	9-2	5	P	-17	106
	05APR2008	LING	10	SD	50	6HCP	2	6	8-11	1	P	+32	91
	20APR2008	GREA	10	SD	40	6HCP	2	13	9-4	7	H	---	100
	28APR2008	LING	10	SW	61	6HCP	2	9	9-0	2	P	-1	103
	05MAY2008	KEMP	10	SD	65	6HCP	8	13	9-0	13	H	+24	89
2 7 LUNAR RIVER	9-2(T)		DAVID PINDER			FERGUS SWEENEY							
	07MAY2007	KEMP	10	SW	54	6HCP	9	12	9-2	7	H	---	92
	11MAY2007	LING	10	SD	---	5HCP	4	11	9-10	1	H	---	101
	23MAY2007	LING	10	SD	40	6HCP	1	9	9-2	9	H	---	99
	12JUL2007	LING	8	SD	56	5HCP	11	11	9-7	2	P	---	81
	21JUL2007	LING	10	SD	---	5HCP	3	10	9-4	11	H	---	102
	31AUG2007	WOLV	8.6	SD	56	6HCP	3	11	9-11	1	P	---	102
	06SEP2007	WOLV	9.5	SD	44	5HCP	8	10	8-11	9	H	---	99
	20SEP2007	WOLV	9.5	SW	50	6HCP	3	11	9-7	10	H	---	94
	05OCT2007	WOLV	9.5	SD	42	5HCP	10	13	8-12	4	H	---	92
3 14 COMPETITOR	9-2(T)		J AKEHURST			KIRSTY MILCZAREK(3)							
	20JAN2007	LING	12	SD	38	5HCP	7	14	8-6	12	P	---	92
	29JAN2007	LING	10	SD	38	6SEL	1	10	9-2	6	P	---	100
	05FEB2007	LING	10	SD	45	6HCP	7	14	9-1	10	P	---	100
	05MAR2007	LING	10	SD	38	6HCP	1	14	9-1	12	H	---	95
	24MAR2007	KEMP	12	SD	52	5HCP	10	11	8-6	2	H	---	89
	30MAR2007	LING	10	SD	42	5SEL	3	8	9-3	3	P	---	98
	05JUN2007	LING	12	SW	---	6HCP	7	13	9-1	8	H	---	95
	30JAN2008	LING	10	SW	64	6HCP	9	14	9-2	5	H	-5	99
	15FEB2008	LING	12	SD	55	6SEL	1	5	9-5	6	L	+38	92
	20FEB2008	LING	10	SD	48	6SEL	2	6	8-12	7	P	+12	102

	19MAR2008	LING	10	SD	36	6SEL	1	9	8-9	5	L	-6	106
	05APR2008	LING	10	SD	50	6HCP	6	6	9-1	4	L	+32	88
	07JUN2008	LING	12	SD	67	6SEL	10	13	9-12	11	P	---	83
4 11 MOONLIGHT FANTASY	Non-Runner												
5 4 PRINCE CHARLEMAGNE	9-1(B)		R M STRONGE				DANE O'NEILL						
	10JAN2007	LING	12	SD	43	4HCP	5	10	8-12	4	H	---	101
	03FEB2007	LING	10	SD	50	4HCP	5	7	9-1	4	H	---	106
	12FEB2007	WOLV	9.5	SD	40	4HCP	4	8	9-0	3	H	---	95
	19FEB2007	LING	10	SD	25	4HCP	4	12	8-9	4	P	---	104
	09MAR2007	LING	10	SD	42	4HCP	6	14	8-13	6	H	---	104
	21MAR2007	KEMP	11	SD	65	4HCP	5	11	8-9	3	H	---	106
	09MAY2007	KEMP	12	SD	45	4HCP	8	11	9-3	10	H	---	86
	03JAN2008	LING	10	SD	38	5HCP	6	9	9-6	5	H	+19	99
	09JAN2008	WOLV	9.5	VS	42	5HCP	3	6	9-0	1	P	-12	100
	09FEB2008	LING	10	SD	71	5HCP	4	8	9-2	6	H	+21	96
	28FEB2008	LING	12	SD	57	4HCP	2	7	8-9	4	H	-1	106
	05MAR2008	LING	12	SD	64	6CLM	5	8	9-11	8	P	+16	95
	19MAR2008	LING	10	SD	36	6SEL	2	9	8-12	7	H	-6	103
6 9 BUNDLE UP	8-13		MRS L J MONGAN				IAN MONGAN						
	16JUN2007	LING	12	SD	---	6MDN	4	6	9-6	5	H	---	94
	11JUN2008	KEMP	11	SD	60	6HCP	5	13	9-1	4	H	+10	92
7 2 THERMIDOR	8-13(B)		LADY HERRIES				RICHARD KINGSCOTE						
	18FEB2008	LING	7	SD	48	5MDN	6	7	9-12	1	H	+4	100
	28MAR2008	LING	8	SD	52	5HCP	6	7	9-1	6	P	+18	94
8 13 GREEN PIRATE	8-13(V)		C R DORE				L P KENIRY						
	13JAN2007	WOLV	8.6	SW	42	6HCP	6	13	8-8	5	H	---	93
	21JAN2007	WOLV	8.6	SD	48	6HCP	8	13	8-13	12	H	---	95
	31JAN2007	LING	8	SD	55	6HCP	2	11	8-6	12	H	---	99
	25JUN2007	WOLV	7.1	SD	56	6HCP	4	12	9-4	6	H	---	94
	02JUL2007	WOLV	8.6	SD	33	6HCP	3	13	9-8	5	H	---	94
	10JUL2007	WOLV	7.1	SD	50	6HCP	2	10	9-4	1	H	---	100
	12AUG2007	WOLV	8.6	SW	56	5HCP	9	13	8-13	7	H	---	93
	30AUG2007	WOLV	7.1	SD	67	6HCP	8	12	9-4	7	H	---	90

	31AUG2007	WOLV	6	SD	56	6HCP	3	13	9-2	8	H	---	97
	08SEP2007	WOLV	8.6	SD	44	6HCP	5	13	9-5	4	H	---	100
	22SEP2007	WOLV	7.1	SD	39	6HCP	2	11	9-0	6	H	---	100
	05OCT2007	WOLV	6	SD	42	6CLM	4	13	8-13	7	H	---	94
	26OCT2007	WOLV	7.1	SD	50	6HCP	4	10	9-6	4	H	---	99
	20DEC2007	WOLV	7.1	VS	42	6HCP	2	11	9-2	5	H	---	101
	26DEC2007	WOLV	7.1	VS	36	6CLM	3	11	8-7	1	H	---	100
	02JAN2008	WOLV	7.1	SW	52	6CLM	7	8	8-11	7	H	-20	91
	20JAN2008	WOLV	7.1	SW	57	6HCP	4	12	8-10	7	H	---	103
	25JAN2008	WOLV	7.1	SD	48	6SEL	1	9	8-12	9	H	-5	95
	09FEB2008	WOLV	8.6	SW	61	6HCP	6	8	9-4	5	P	+30	89
	13FEB2008	LING	8	SD	31	6CLM	3	7	8-9	6	H	+23	95
	15FEB2008	WOLV	7.1	SW	58	6HCP	3	12	8-13	3	H	-19	101
	03MAR2008	WOLV	6	SW	48	6CLM	1	8	8-9	4	H	+5	99
	28MAR2008	WOLV	6	SD	61	6HCP	6	7	9-2	4	P	+24	82
	19MAY2008	WOLV	6	SD	47	6HCP	8	13	9-0	10	H	-7	89
	06JUN2008	WOLV	7.1	SD	56	6HCP	8	12	9-1	3	H	-32	93
	14JUN2008	LING	8	SD	61	6HCP	5	9	9-3	5	H	---	97
9 10 CONVALLARIA	8-13(T)		G WRAGG				WILLIAM BUICK						
	04MAY2007	LING	7	SW	44	5MDN	6	10	9-5	7	H	---	96
	18SEP2007	LING	8	SD	48	5MDN	7	11	9-2	3	H	---	92
	16NOV2007	LING	7	SD	52	6HCP	5	14	9-0	13	H	---	99
	26NOV2007	LING	7	SD	46	6HCP	3	11	9-0	10	H	---	97
	18JAN2008	KEMP	8	SD	57	6HCP	2	12	9-0	10	H	+5	99
	30JAN2008	LING	8	SW	64	6HCP	8	11	9-1	9	H	+20	98
	25FEB2008	WOLV	8.6	SW	67	6HCP	2	7	9-0	2	H	0	100
	26MAR2008	WOLV	9.5	SD	47	6HCP	8	10	9-2	1	H	-19	93
	05MAY2008	KEMP	10	SD	65	6HCP	9	13	8-10	10	P	+24	89
10 6 MIX N MATCH	8-10		R M STRONGE				TRAVIS BLOCK(3)						
	31MAY2007	LING	6	SD	36	5MDN	8	10	9-0	3	H	---	87
	02FEB2008	KEMP	7	SD	38	6HCP	3	10	8-9	6	H	-8	95
	10FEB2008	KEMP	8	SD	52	6HCP	2	10	8-7	4	H	---	104
	16FEB2008	KEMP	8	SD	45	6HCP	10	12	8-8	1	H	+9	86

11 1 BOWL OF CHERRIES	8-6(V)		I A WOOD				JAMES DOYLE						
	16JUN2007	LING	10	SD	---	5HCP	8	8	8-8	1	P	---	92
	23JUN2007	LING	10	SW	---	6HCP	2	12	8-12	3	P	---	102
	27JUN2007	KEMP	8	SD	60	6HCP	9	13	8-12	3	H	---	94
	29JUN2007	WOLV	8.6	SD	61	6HCP	7	13	11-1	11	P	---	89
	11JUL2007	KEMP	8	SD	48	6HCP	6	14	9-2	9	H	---	96
	04AUG2007	LING	10	SD	---	6HCP	10	14	9-2	12	L	---	92
	29AUG2007	KEMP	10	FT	63	6HCP	4	14	8-13	11	P	---	95
	05SEP2007	KEMP	11	FT	55	6HCP	5	13	8-13	14	P	---	97
	14SEP2007	WOLV	12.2	FT	61	6HCP	8	12	8-13	10	P	---	94
	29OCT2007	KEMP	10	SD	57	6HCP	7	13	8-13	8	H	---	96
	31OCT2007	KEMP	10	FT	33	6HCP	1	12	8-10	4	H	---	106
	07NOV2007	KEMP	10	FT	36	6HCP	4	14	9-2	10	P	---	100
	03DEC2007	WOLV	9.5	SW	38	6HCP	9	13	8-12	10	H	---	87
	09DEC2007	LING	10	SD	50	6HCP	6	11	9-1	12	P	---	90
	12DEC2007	KEMP	10	SD	62	6HCP	7	13	9-1	7	H	---	100
	02APR2008	KEMP	11	SW	36	6HCP	5	11	8-10	3	H	+11	98
	08APR2008	SOUT	11	VS	50	6HCP	6	14	9-0	13	H	-12	96
	10APR2008	KEMP	10	SD	58	6HCP	4	11	8-11	10	H	---	96
	08JUN2008	SOUT	8	SD	67	6HCP	9	14	8-8	11	H	---	87
12 3 FILM QUEEN	8-6		MRS L J MONGAN				PAUL DOE						
	23JAN2008	KEMP	7	SD	57	5MDN	8	14	9-1	13	H	+11	85
	02FEB2008	LING	7	SD	76	5MDN	4	6	9-1	3	H	+32	76
	16FEB2008	LING	6	SD	52	5MDN	2	10	9-0	10	H	+15	85
	26FEB2008	LING	7	SD	55	6HCP	9	10	8-11	8	H	+10	89
	14MAR2008	LING	7	SW	45	5MDN	10	11	9-4	6	H	+18	85
	28MAR2008	KEMP	8	SD	45	6CLM	5	9	8-4	9	H	+76	80
	05APR2008	KEMP	7	SW	33	6HCP	5	14	8-10	8	P	-2	93
	08MAY2008	GREA	10	VS	61	6CLF	16	16	9-8	16	H	---	45
13 8 TENEMENT	8-6		JAMIE POULTON				SIMON WHITWORTH						
	22JAN2007	KEMP	7	SD	27	5MDN	8	14	9-3	8	P	---	86
	29JAN2007	LING	8	SD	38	5MDN	8	11	9-3	10	P	---	89
	10FEB2007	LING	8	SD	48	6CLM	2	7	9-5	5	P	---	82
	14JUN2008	LING	10	SD	61	5HCP	4	11	8-2	10	H	---	101
14 12 REVEUR	*Non-Runner*												

Draw and Front Runner Statistics for 10 Furlongs during past 12 months

Draw Pos	Winners	%	
Low	28	29	
Middle	36	38	
High	32	33	
DRAW NO	**RUNNERS**	**WINNERS**	**%**
1	94	7	7
2	91	12	13
3	90	9	10
4	91	10	11
5	89	16	18
6	88	9	10
7	80	7	9
8	70	5	7
9	50	5	10
10	46	4	9
11	43	4	9
12	35	3	9
13	24	2	8
14	20	3	15

Front Runners Impact Value = 0.9
Prominent Horses Impact Value = 1.3
Hold Up Horses Impact Value = 0.8

Within the past couple of seasons both Kings Topic and Prince Charlemagne had earned decent speed figures (well in excess of the class par) on more than one occasion over this course and distance. They clincher which made them both worth a punt was that they were also well drawn.

Kings Topic went off the 5-2 favourite and ran creditably to finish fifth, but the money was landed when Prince Charlemagne came home in front at a very tasty 14-1. This wasn't the greatest contest in the world, but no matter when the information you have at hand pays dividends.

The following contest from the Kempton meeting on June 25th 2008 demonstrates perfectly the value of knowing horses' running styles alongside the inverted draw statistics.

7-50 SALLY NICHOLLS (I AM BACK) HANDICAP (0-80) (4) 3YO+

6 furlongs **Par = 105** **12 runners**

1 5 DIRICULOUS	10-0		T G MILLS			JIM CROWLEY							
	31OCT2007	KEMP	6	FT	33	6MDN	4	5	9-3	4	H	---	92
	28NOV2007	KEMP	6	SD	48	6MDN	1	12	9-3	10	P	---	102
	10JAN2008	SOUT	6	SD	83	6HCP	1	9	9-3	3	L	+8	101
	17JAN2008	SOUT	5	SD	67	6HCP	5	13	9-9	10	H	-2	94
	28FEB2008	KEMP	6	SD	69	5HCP	1	12	8-8	9	P	-10	105
	02APR2008	KEMP	6	SW	36	4HCP	2	10	8-12	7	H	+1	104
	08APR2008	SOUT	6	VS	50	5HCP	1	8	9-1	7	P	-25	102
	20MAY2008	SOUT	6	SD	37	4HCP	1	11	9-1	8	H	-34	100
2 2 KYLE *Non-Runner*													
3 11 WHISKEY JUNCTION	9-11		A M BALDING			L P KENIRY							
	07MAY2007	KEMP	5	SW	54	4HCP	2	7	8-5	5	P	---	93
4 12 DVINSKY	9-9(B)		P HOWLING			JIMMY QUINN							
	08JAN2007	WOLV	6	SD	42	5HCP	1	8	8-12	1	L	---	96
	28JAN2007	KEMP	6	SD	40	5HCP	9	11	9-1	6	H	---	94
	05FEB2007	WOLV	7.1	SD	25	5HCP	7	9	9-0	6	P	---	92
	21FEB2007	WOLV	6	SD	35	5HCP	10	13	8-8	13	H	---	94
	05MAR2007	WOLV	6	SD	47	5HCP	3	13	8-11	10	P	---	103
	27MAR2007	WOLV	6	FT	45	5HCP	3	13	9-1	5	P	---	103
	25APR2007	KEMP	6	SD	55	5HCP	1	11	8-10	9	L	---	104
	27JUN2007	KEMP	6	SD	60	4HCP	2	12	9-6	4	P	---	99
	17JUL2007	KEMP	6	FT	64	4HCP	3	11	9-1	3	P	---	98
	30JUL2007	WOLV	6	SD	56	5HCP	5	13	9-10	12	P	---	97
	04SEP2007	LING	6	SD	31	4HCP	5	10	8-12	5	P	---	97
	22SEP2007	WOLV	6	SD	39	5HCP	2	13	9-1	8	P	---	100
	11OCT2007	KEMP	6	FT	50	5HCP	3	11	9-2	11	L	---	99
	17OCT2007	LING	6	SW	50	4HCP	4	10	8-10	11	P	---	101
	30OCT2007	WOLV	7.1	SD	62	5HCP	5	11	9-3	3	L	---	92
	14NOV2007	WOLV	6	SW	43	5HCP	9	13	9-4	1	P	---	95
	26NOV2007	LING	6	SD	46	5HCP	7	12	9-0	5	P	---	101
	05DEC2007	KEMP	6	SD	67	5HCP	1	9	9-0	8	L	---	102

		16DEC2007	KEMP	7	FT	21	5HCP	6	14	9-4	4	P	---	98
		05JAN2008	KEMP	6	SD	55	5HCP	10	10	8-13	4	L	---	95
		15JAN2008	KEMP	6	SW	38	5HCP	6	9	9-3	5	P	---	104
		28JAN2008	KEMP	6	SD	29	5HCP	4	9	9-2	7	P	---	103
		04FEB2008	LING	7	SD	67	6CLM	3	6	9-6	6	P	-6	99
		13FEB2008	KEMP	6	SW	41	5HCP	1	11	9-2	2	P	+3	105
		03MAR2008	LING	7	SD	53	5HCP	4	9	8-13	2	L	+21	101
		15MAR2008	KEMP	6	FT	50	5HCP	2	11	9-4	4	P	-2	105
		02APR2008	LING	7	SD	50	5HCP	1	7	9-4	1	L	+15	103
		17APR2008	KEMP	6	SD	36	4HCP	3	8	8-13	7	L	+7	107
		23APR2008	KEMP	6	SD	47	4HCP	2	6	8-13	1	L	+7	102
5 1 CHINA CHERUB	*Non-Runner*													
6 6 STEEL CITY BOY		9-6			MISS A STOKELL			ANN STOKELL(5)						
		08MAY2007	SOUT	5	FT	---	5HCP	1	11	8-12	4	P	---	103
		02NOV2007	WOLV	6	SD	72	5HCP	8	10	9-4	8	P	---	94
		28APR2008	SOUT	5	SW	17	4HCP	4	6	8-7	3	L	-12	104
		20MAY2008	SOUT	6	SD	37	4HCP	3	11	8-7	10	P	-34	98
7 3 RABBIT FIGHTER	*Non-Runner*													
8 10 SPOOF MASTER		9-4			C R DORE			LIAM JONES						
		07APR2007	KEMP	6	SD	62	2HCP	5	8	8-13	7	L	---	99
		30NOV2007	LING	6	SD	62	4HCP	9	10	9-1	1	H	---	93
		11DEC2007	SOUT	5	VS	---	3HCP	9	10	8-7	7	H	---	92
		01JAN2008	SOUT	6	SD	64	5HCP	7	8	8-13	1	P	---	93
		07FEB2008	SOUT	5	SD	67	5HCP	2	6	9-1	1	P	0	103
		14FEB2008	SOUT	5	VS	83	5HCP	2	8	9-4	2	L	-2	99
		28FEB2008	KEMP	5	SD	69	5CLM	2	6	8-12	1	P	+13	100
		26MAR2008	WOLV	5.1	SD	47	4HCP	4	7	8-7	5	P	+1	99
		07APR2008	KEMP	5	SD	62	5HCP	2	7	9-4	7	L	+7	102
		06MAY2008	SOUT	5	SW	83	5HCP	3	10	9-2	1	P	-6	100
		04JUN2008	KEMP	5	SD	57	5HCP	6	12	10-0	10	L	-5	94
9 4 ADANTINO		9-2(B)			B R MILLMAN			T G MCLAUGHLIN						
		10FEB2007	LING	6	SD	48	4HCP	4	12	8-9	2	H	---	106
		27FEB2007	LING	6	SD	43	4HCP	10	11	8-11	3	H	---	99

	17MAR2007	LING	7	SD	38	4HCP	4	13	8-12	8	H	---	100
	30AUG2007	LING	6	SD	63	4HCP	8	12	9-2	8	H	---	97
	08SEP2007	KEMP	7	SD	33	4HCP	10	13	9-3	3	L	---	96
	22SEP2007	WOLV	6	SD	39	5HCP	7	13	8-12	4	H	---	99
10 7 HUCKING HILL	8-12(B)		J R BEST			TED DURCAN							
	18JUL2007	LING	6	SD	---	5HCP	4	9	9-9	8	H	---	95
	23AUG2007	LING	6	FT	47	5HCP	4	12	9-7	8	H	---	96
	27AUG2007	KEMP	6	SD	67	5HCP	2	10	9-1	4	H	---	99
	24SEP2007	KEMP	6	SD	54	4HCP	2	10	8-3	10	H	---	97
	11OCT2007	KEMP	6	FT	50	5HCP	8	11	9-0	3	H	---	94
	15OCT2007	LING	7	SW	46	5HCP	10	14	8-13	4	H	---	95
	30NOV2007	LING	6	SD	62	4HCP	10	10	8-5	2	H	---	91
	05JAN2008	KEMP	6	SD	55	5HCP	6	10	8-11	10	P	---	100
	09JAN2008	LING	6	SD	55	5HCP	3	6	8-12	3	L	---	95
	13JAN2008	KEMP	6	SD	43	5HCP	5	10	8-7	2	P	---	100
	15JAN2008	KEMP	6	SW	38	5HCP	3	9	9-0	9	P	---	106
	28JAN2008	KEMP	6	SD	29	5HCP	5	9	9-0	4	H	---	103
	02FEB2008	KEMP	6	SD	38	5HCP	5	8	9-0	5	H	+24	104
	28FEB2008	KEMP	6	SD	69	5HCP	4	12	8-8	11	H	-10	102
	05MAY2008	KEMP	10	SD	65	6HCP	11	13	9-4	7	P	+24	80
11 8 REPLICATOR	8-12		PAT EDDERY			PAT DOBBS							
	14APR2007	LING	5	SW	50	5MDN	7	7	8-9	3	H	---	65
	07MAY2007	KEMP	5	SW	54	4MDN	4	7	9-3	7	P	---	79
	11JUL2007	LING	6	SW	58	6MDN	5	12	8-12	12	P	---	88
	08AUG2007	KEMP	6	SD	64	4HCP	2	9	8-3	7	L	---	89
	18SEP2007	LING	6	SD	48	5HCP	4	8	9-0	8	P	---	89
	28MAY2008	GREA	5	SD	58	5HCP	3	9	9-2	10	P	---	94
	11JUN2008	KEMP	6	SD	60	4HCP	4	11	8-13	8	P	+1	93
12 9 SHERIFF'S SILK	8-11(B)		G D BLAKE			OSCAR URBINA							
	11JAN2007	SOUT	7	SW	67	6SEL	1	11	9-4	6	L	---	89
	20FEB2007	SOUT	7	FT	65	5HCP	3	7	8-10	7	P	---	103
	13MAR2007	SOUT	6	SW	52	5HCP	2	7	9-2	4	L	---	106
	22APR2007	SOUT	6	SD	57	4HCP	2	9	8-11	6	P	---	98
	24MAY2007	SOUT	7	SD	60	4HCP	5	8	8-13	8	L	---	100
	24APR2008	SOUT	7	SD	69	5HCP	4	4	9-4	1	P	+7	88

Draw and Front Runner Statistics for 6 Furlongs during past 12 months

Draw Pos	Winners	%	
Low	25	23	
Middle	24	22	
High	62	56	
FROM RAIL	**RUNNERS**	**WINNERS**	**%**
1	105	29	28
2	102	20	20
3	101	8	8
4	105	8	8
5	103	9	9
6	97	9	9
7	91	10	11
8	85	5	6
9	71	4	6
10	53	1	2
11	39	6	15
12	19	2	11

Front Runners Impact Value = 1.9
Prominent Horses Impact Value = 1.2
Hold Up Horses Impact Value = 0.6

Everything pointed to a complete early burn-up against the inside rail. Whiskey Junction had shown plenty of early speed in winning his last two starts on turf and the running styles of the other three highest-drawn horses are there for all to see. Not all of them could lead early of course, and the deciding factor would have to be the winning records of the two berths closest to the inside rail occupied by Dvinsky and Whiskey Junction. Which one of them would be able to get to the front early and therefore become the most likely winner if the Front-Runners Impact Value is taken into account alongside the draw statistics?

Of course the bet here was the Exacta even though the aforementioned pair were amongst the top three in the betting. As it turned out it was Whiskey Junction who was able to beat Dvinsky to the first bend and the pair dominated throughout. Despite Dvinsky trying hard to get back at the leader, he could never quite get there and Whiskey Junction held on by a neck to win at 3-1 from Dvinsky at 4-1. Incidentally, Spoof Master was held up and never figured, but Sheriff's Silk finished strongly from well off the pace to finish third at 25-1.

The Exacta paid £12 to a £1 stake which is not exactly a huge amount, but still worthwhile when you have the evidence in front of you like this.

I particularly liked one of the horses in this race at Wolverhampton on July 1st 2008.

5-00 TRY BETDAQ FOR AN EXCHANGE HANDICAP (0-60) (6) 3YO+

8.6 furlongs **Par = 101** **13 runners**

1 5 BENTLEY	10-3		J G GIVEN				T P QUEALLY						
	04JAN2007	WOLV	6	SD	60	5HCP	3	7	7-9	4	L	---	96
	11JAN2007	SOUT	5	SD	---	6HCP	2	6	8-6	6	H	---	98
	21JAN2007	WOLV	6	SD	48	6HCP	1	9	8-12	2	L	---	90
	29JAN2007	LING	6	SD	38	5HCP	4	9	8-4	3	L	---	94
	06FEB2007	SOUT	5	SW	---	6HCP	1	9	9-3	7	P	---	95
	09FEB2007	WOLV	5.1	SW	---	6HCP	1	8	9-2	9	P	---	96
	19FEB2007	LING	5	SD	25	5HCP	6	10	9-2	5	P	---	93
	11JUN2007	WOLV	6	SD	44	4HCP	2	7	8-0	5	H	---	95
	14JUN2007	LING	6	SW	44	5HCP	4	9	9-0	5	P	---	95
	02JUL2007	WOLV	6	SD	33	6HCP	12	13	9-4	6	H	---	88
	27AUG2007	KEMP	6	SD	67	5HCP	8	10	8-4	2	H	---	85
	11DEC2007	SOUT	6	SD	39	6HCP	3	10	9-0	9	H	---	93
	31DEC2007	WOLV	5.1	SW	62	6HCP	2	10	9-0	9	P	---	96
	07JAN2008	SOUT	6	SD	57	6HCP	2	7	8-6	4	P	---	99
	12JAN2008	WOLV	5.1	SW	56	6HCP	2	8	8-13	1	P	-5	102
	17JAN2008	SOUT	5	SD	67	6HCP	2	13	8-12	1	P	-2	99
	31JAN2008	SOUT	5	SD	67	6HCP	5	9	8-6	7	P	-12	95
	10FEB2008	KEMP	6	SD	52	6HCP	10	12	8-7	8	P	---	98
	15FEB2008	WOLV	6	SW	58	6HCP	8	9	8-12	8	H	-1	90
	24FEB2008	KEMP	5	SW	52	6HCP	4	12	8-13	12	P	---	98
	03MAR2008	LING	5	SD	53	6HCP	6	8	8-13	5	P	-4	97
	22MAR2008	WOLV	5.1	SD	56	6HCP	3	7	8-8	2	H	-5	95
	26MAR2008	WOLV	5.1	SD	47	6HCP	2	9	8-9	1	L	+1	98
	04APR2008	SOUT	5	FT	75	6HCP	2	13	8-10	8	P	---	99
	29APR2008	WOLV	5.1	SD	56	6HCP	4	9	9-0	12	L	-4	96
	23JUN2008	WOLV	7.1	SW	42	6HCP	1	10	9-4	8	P	-3	101
2 1 KANSAS GOLD	10-0		J MACKIE				ROYSTON FFRENCH						
	14MAY2007	SOUT	6	FT	55	5HCP	11	14	8-11	14	H	---	82
	20SEP2007	WOLV	9.5	SW	50	6HCP	1	11	9-3	8	P	---	96
	05OCT2007	WOLV	9.5	SD	42	5HCP	3	13	8-12	12	P	---	100
	12OCT2007	WOLV	9.5	SW	36	5HCP	1	11	8-12	10	P	---	94
	20OCT2007	WOLV	9.5	SD	47	6HCP	10	12	9-9	10	H	---	92

	30NOV2007	WOLV	9.5	SW	64	5HCP	5	11	8-13	10	P	---	99
	17DEC2007	WOLV	9.5	VS	50	5HCP	6	7	8-12	7	L	---	83
	11JAN2008	WOLV	9.5	VS	55	6HCP	6	9	9-5	2	P	+12	93
3 10 MONTEMAYORPRINCESS	9-9(P)		D HAYDN JONES				TED DURCAN						
	25JUN2007	WOLV	5.1	SD	56	5HCP	10	10	8-4	4	H	---	77
	03DEC2007	WOLV	9.5	SW	38	6HCP	4	13	8-5	12	P	---	93
	01JAN2008	SOUT	8	SD	64	6HCP	1	10	8-2	10	P	---	108
	22JAN2008	SOUT	8	FT	67	6HCP	4	9	8-6	6	P	-9	94
	17MAR2008	WOLV	8.6	SD	55	6HCP	1	8	8-11	3	P	+1	99
	08APR2008	SOUT	8	VS	50	6HCP	5	9	8-13	9	P	-2	84
4 6 FAIR SAILING	9-8		J W HILLS				WILLIAM BUICK						
	14SEP2007	WOLV	8.6	FT	61	6MDN	4	13	8-12	8	P	---	92
	15OCT2007	LING	8	SW	46	5MDN	4	9	8-12	1	P	---	91
	06NOV2007	LING	8	SD	38	6MDN	4	8	8-12	4	P	---	93
5 12 PASSATO	9-7(B)		R A HARRIS				L P KENIRY						
	02JAN2008	WOLV	8.6	SW	52	5HCP	7	9	8-8	2	H	+7	96
	16JAN2008	KEMP	8	SD	40	5HCP	9	12	8-13	10	P	+10	92
	11FEB2008	WOLV	9.5	SD	60	6HCP	7	13	8-10	10	H	-6	94
6 2 WILLIE EVER	9-6		B ELLISON				J-P GUILLAMBERT						
	27DEC2007	WOLV	8.6	VS	64	6MDN	6	7	9-3	6	H	---	89
	21JAN2008	WOLV	9.5	SW	55	6HCP	4	9	8-9	8	L	+6	95
	25APR2008	WOLV	7.1	SW	53	6HCP	2	12	8-8	7	P	-8	100
7 3 KIMONO MY HOUSE	9-4		J G GIVEN				PAT COSGRAVE						
	14SEP2007	WOLV	12.2	FT	61	6HCP	7	11	8-7	11	H	---	92
8 13 RAMBLING SOCKS	9-4(P)		S R BOWRING				STEPHEN DONOHOE						
	12JAN2007	WOLV	7.1	VS	38	5MDN	7	11	9-8	8	P	---	88
	27JAN2007	WOLV	8.6	SW	22	7CLF	7	11	8-13	12	P	---	85
	05MAR2007	WOLV	9.5	SD	47	5CLM	12	12	8-2	3	H	---	73
	13MAR2008	SOUT	8	SW	50	6SEL	2	8	9-8	8	P	+19	92
	22MAR2008	WOLV	6	SD	56	7CLF	1	13	9-4	1	P	-8	97
	29APR2008	SOUT	8	SW	47	6CLM	3	7	7-13	7	P	+2	92
	03JUN2008	SOUT	6	FT	57	6HCP	4	11	8-13	2	H	-17	91

	08JUN2008	SOUT	6	SD	67	6HCP	4	12	8-10	5	P	---	96
9 11 ARDENT PRINCE	9-2		A J MCCABE			SEB SANDERS							
	20AUG2007	WOLV	13.9	SD	36	6HCP	2	13	9-2	4	L	---	101
	07SEP2007	KEMP	16	FT	67	5HCP	8	13	8-7	5	L	---	96
	13SEP2007	WOLV	13.9	SD	50	6HCP	10	12	9-6	7	H	---	91
	17OCT2007	KEMP	12	SD	36	6HCP	11	14	9-4	3	L	---	93
	08JUN2008	SOUT	8	SD	67	6HCP	13	14	8-0	13	P	---	72
10 8 THE GRAIG	9-2		J R HOLT			DAVID KINSELLA							
	25APR2007	KEMP	11	SD	55	6MDN	5	9	8-8	9	H	---	77
	08MAY2007	SOUT	11	SW	53	5MDN	11	13	8-10	13	H	---	65
	09JUN2007	LING	10	SD	---	6HCP	11	14	8-7	14	H	---	92
	17JUL2007	KEMP	12	FT	64	6HCP	12	13	8-9	1	H	---	86
	05APR2008	KEMP	8	SW	33	6HCP	5	14	8-3	5	P	0	97
	25APR2008	WOLV	7.1	SW	53	6HCP	9	2	8-2	2	P	-8	96
	19JUN2008	GREA	8	SD	50	6HCP	4	15	8-5	15	P	---	94
11 7 WOGAN'S SISTER	9-1		I A WOOD			SHANE KELLY							
	23OCT2007	LING	7	SD	50	5MDN	9	10	8-8	3	P	---	84
	07NOV2007	KEMP	7	FT	36	6MDN	8	12	8-8	6	P	---	83
	13MAR2008	WOLV	8.6	SW	58	6HCP	1	11	8-10	1	P	-5	94
12 4 SOLO RIVER	9-0		P J MAKIN			FERGUS SWEENEY							
	13OCT2007	KEMP	6	FT	33	5MDN	4	9	8-4	4	H	---	87
	24OCT2007	KEMP	5	FT	69	6MDN	7	10	8-4	4	P	---	83
	09APR2008	KEMP	7	SD	50	6HCP	6	11	8-13	7	H	-3	84
	28APR2008	LING	6	SW	61	6HCP	6	10	9-0	6	H	0	97
	19MAY2008	WOLV	7.1	SD	47	6HCP	3	11	8-13	2	P	-17	87
	07JUN2008	LING	10	SD	67	6HCP	4	13	8-11	10	P	---	98
13 9 SUPPORTING ROLE	8-11(B)		E S MCMAHON			RICHARD MULLEN							
	22MAR2008	WOLV	9.5	SD	56	6HCP	2	11	9-2	5	P	+4	94
	12APR2008	WOLV	9.5	SW	50	6HCP	5	8	9-3	6	P	+11	90

Draw and Front Runner Statistics for 8.6 Furlongs during past 12 months

Draw Pos	Winners	%	
Low	68	42	
Middle	45	28	
High	50	31	
DRAW NO	**RUNNERS**	**WINNERS**	**%**
1	153	22	14
2	152	23	15
3	148	22	15
4	154	15	10
5	149	17	11
6	143	11	8
7	139	15	11
8	118	14	12
9	94	2	2
10	80	7	9
11	63	6	10
12	48	5	10
13	43	4	9

Front Runners Impact Value = 1.2
Prominent Horses Impact Value = 1.2
Hold Up Horses Impact Value = 0.9

Willie Ever had just the sort of credentials I like. There was no doubt over his ability to act on the surface nor was the trip likely to be a problem. He also had a great draw in stall two, but although he had been beaten on turf since his last start on sand, that was over six furlongs and that trip was far too short for him. What particularly attracted me to him was that each of the speed figures he had earned in his three outings on sand was higher than the previous one. Even a modest improvement on his most recent figure would take him over the class par for this contest and probably into the winner's enclosure.

A lot of people must have agreed with me because he was backed in to 2-1 favourite having touched 10-3, but he gave his supporters few anxious moments and flew home to record a very comfortable two-and-three-quarter-length victory.

All of these example races may smack of being clever after the event, but I know full well that without the tools at my disposal, whether purchased from a third party or of my own creation, I wouldn't have had the knowledge to dissect these races in quite the same way, nor to the same effect. The rewards are there if you are prepared to put the work in.

JOURNAL

ORIGINALLY I wasn't going to include a Journal in this publication, or a diary as I called it in my previous books. I didn't want this to be a rehash of Gold From The Sands I and II, I wanted this one to be different in as many ways as possible. In other words I wanted this to read and sound fresh, but then I did some thinking.

Could I really expect anyone to read this book and then go out and perhaps apply one or two of the recommendations within it if the author wasn't willing to put his hand in his pocket and risk his own money? Isn't it true that the best way to demonstrate any theory or idea is to do it in the field? Whilst the opinions and ideas contained within this publication may be familiar to some people – especially if they have read any of the 200-odd King Of The Sand columns I've written over the past ten years – for others it may be the first time they have had the chance to digest any of this. So the bottom line is that I felt it was my duty to go out there and get my hands dirty, to duck the bullets flying over my head, and to show that everything I have written about up to now works when it really matters.

I also realised that even though I have done this sort of thing before in previous books, this would be a fresh challenge and most of the horses I would see in the coming months wouldn't be the same ones that I backed in 2000 or 2005 (I say 'most' because a few All-Weather stalwarts do tend to re-emerge Lazarus-like year in year out) and, even more importantly, at this stage I don't know if I'm going to end up in profit or not when this betting period ends in the summer. Things have worked out okay when I've done this sort of thing in the past and I'm confident they will again, but nothing is guaranteed.

In Gold From The Sand II my level stake for the diary was £40 per horse, this time it will be £50 per horse – such is the effect of inflation – and I may also use exactas when appropriate. One thing I have learnt over the years is that when you have statistics available to you that have exposed a track and/or draw bias, and after careful study you are in a position to take advantage of it, the win bet is all very well and good, but the exacta can really press the advantage home.

So here goes. This is what happened on my visits to the All-Weather tracks in the spring and summer of 2008.

Note: For each meeting shown in this diary, I would have used the full detailed version of my racecards shown earlier in the book as my main source of information. Unfortunately, due to lack of space, it is impossible to reproduce them in this section and hence the more abbreviated Private Handicap-style cards are shown for the races I actually invested in.

Saturday 15 March – Lingfield

This was the day after the great clash between Denman and Kauto Star in the Cheltenham Gold Cup and the racing press were still buzzing about it. Having said that I came across four racing professionals during the course of the afternoon – three well-known journalist/commentators plus a representative of one of the big three bookmaking firms – and they all understandably looked a little jaded after a few very tough and demanding days at Prestbury Park. Still, being professional to the core they soldiered on through another eight-race programme on the Lingfield Polytrack and listening to them you would never have known how demanding their schedule had been.

For me, All-Weather meetings didn't get any better than this. It was Winter Derby Day and quite apart from the Group 3 highlight itself there were also a couple of Listed races and some tasty handicaps to get my teeth into.

The meeting started with a 14-horse maiden for three-year-olds over a mile and two furlongs and if I wanted to start this journal off with a bit of drama then I could hardly have chosen a better race. Unfortunately it wasn't drama in a positive sense.

1-40 LIFESTYLESERVICES.UK.COM MAIDEN STAKES (4) 3YO

10 furlongs **Par = 100** **14 runners**

AW FORM	NAME	BEST	LAST SIX	
3	AZABU JUBAN	100S11	100S11	
52-	NO TO TRIDENT	94W08	84W08	94W08
7-3	NOWZDETIME	93L10	84L07	93L10
57	FAIRFIELD FLAME	90K08	90L08	90K08
0	EMSHABB	83K08	83K08	
0-8	WHENINEEDYOU	73W08	73L07	73W08
0-U	MALT EMPRESS	71L07	71L07	0L10
0-	AMPHIBALUS	54K08	54K08	
	SUMMER WINDS			
	MUMS THE BEST			
	PRINCESS LOMI			
	BARRING DECREE			
	VICE CONSUL			
	EVERY WHISPER			

Although some big stables were represented in this race, there was also a lot of dead wood and because of that I became interested in the highest-rated horse on my card, Azabu Juban. As you can see, on my figures the filly was fast enough to win a race like this, at least that is what she had shown on Fibresand and I was aware that she might not be able to do so on this faster Polytrack. Another potential negative was that I had never heard of the 7lb claimer due to ride her – Natalie Jankiewicz – who it transpired was having only her third ride, but all of this was factored into the price and at 25-1 I felt she was worth a punt in a race where very few had shown any worthwhile form, whilst many of the others lacked experience.

I don't bet to place very often, but this was one of those occasions when that was the right sort of bet, so I invested £25 each-way

I didn't expect what followed, however. As soon as the gates opened, Azabu Juban was put to sleep at the back of the field and she trailed her rivals for most of the journey before running on past a few beaten horses to finish tenth, with her rider doing very little. My initial feeling was that I hadn't got a run for my money and although I am not a conspiracy theorist and try to keep talking through my pocket to a minimum, this was an occasion when there was a small murmur emanating from my wallet. The comment-in-running in the *Racing Post* the following day summed up the performance perfectly.

dwelt: restrained towards rear: ran on final 2f: can do better

It was no surprise when a little time later, the filly's trainer Jonathan Jay was asked to go to the Stewards' Room. "Good", I thought. At least they were going to have a look into the ride. I kept my ears open as the afternoon progressed, so that I could hear the outcome of the enquiry when it was announced.

When it eventually came, it was really quite fascinating to listen to. The rider herself had stated that she was meant to make the running aboard Azabu Juban, but following a slow break the filly was unable to take closer order. The killer blow for Ms Jankiewicz, however, was that even the trainer stated that he wasn't satisfied with the ride and that she should have made more of an effort despite her inexperience. As a result the rider got a 14-day ban for making insufficient effort, whilst Jonathan Jay was exonerated.

When I first heard all this, my first reaction was one of justice being done, but very quickly I regretted feeling that way. The 14-day ban given to the jockey was mainly symbolic, as I doubt that Ms Jankiewicz would have had that many rides in the 14-day period she was going to miss. Had it been a senior jockey involved then the consequences would have been very different, but it still seemed strange to me that a rider with so little experience should take it on herself to canter a horse around one circuit of the Lingfield Polytrack and not put in any effort. Perhaps due to her inexperience she really wasn't capable of giving the horse any more assistance than she did, but either way I soon felt genuinely sorry for her. My feeling on the whole affair was, and still is, one of unease rather than anger.

In any case the fact of the matter was that my journal had started off with me backing a non-trier.

The second race on the card, a Class 2 handicap over seven furlongs, just had to work out better and in quite a valuable race like this I was certain that I would at least get a run for my money.

2-10 DIAMONDRACING.CO.UK HANDICAP (0-100) (2) 4YO+								
7 furlongs		**Par = 109**				**8 runners**		
AW FORM	**NAME**	**BEST**			**LAST SIX**			
1/321-34	VORTEX	114W08	104L07	107L08	104L07	103L07	90L08	101L08
04916-	GALLANTRY	113L08	101L08	105K08	105W07	100W08	107K08	113L08
35-38478	QADAR	113L06	103L06	97L06	102L06	105K06	105L07	99W06
563-1133	MONKEY GLAS	112L08	99L08	106W07	101L08	105K08	112L08	104L08
111-4441	ATLANTIC STORY	111L08	105K08	99L08	97L10	106K08	111L08	110L07
/27231-2	CEREMONIAL JADE	110L06	109L08	105K07	98L07	104L07	110L06	110L06
-2411106	BAZROY	109L07	104L07	104K06	102K08	109L07	106L08	102W06
043/093-	PURUS	103K07	96K07	103K07	98W07	93S07	96K07	100L07

The only problem was that there were several contenders. Gallantry, Monkey Glas and Atlantic Story had met here quite recently with Gallantry faring best, whilst Ceremonial Jade had only just been beaten by Excusez Moi (who was due to run later on the card) in a

Listed race here last time. In the end I narrowed it down to two, and had £50 on Gallantry at 7-1 and the same amount on Ceremonial Jade at 100-30.

Gallantry ran his race and had every chance, but Ceremonial Jade gave himself too much to do despite the longer trip. They ended up finishing second and fourth respectively behind Atlantic Story.

The third race on the card was the Listed Spring Cup for three-year-olds over seven furlongs. This is a race I can never quite suss as it features several decent juveniles from last season with little sand experience against a few with more experience of sand, but without quite the same class.

I didn't need much convincing to swerve this race so didn't get involved, but there was no way I was going to swerve the Winter Derby itself.

3-15 BETDIRECT.COM WINTER DERBY (3RD LEG OF THE EUROPEAN ALL WEATHER SERIES) (GROUP 3) (1) 4YO+

10 furlongs **Par = 113** **14 runners**

AW FORM	NAME	BEST	LAST SIX					
1-253128	ALFRESCO	113L08	105K08	102W08	106K08	113L08	104L08	99W08
822512-0	GRAND PASSION	112L10	108L10	102K10	109K12	104L10	97L10	104L10
1-1	SILVER PIVOTAL	109L10					100W08	109L10
712311-2	EVIDENT PRIDE	109K08	98K08	109K08	103K08	107L10	108K10	97L10
544252-2	ORCHARD SUPREME	108W08	107L08	106K08	100K07	98L08	98L08	102L08
10-	HATTAN	108W08					108W08	106L10
/31-1423	BAYLINI	107L10	107L10	103L10	107L10	102L10	99L10	107L10
1-2	DUBAI'S TOUCH	107L10					104K08	107L10
1/7-	GREAT HAWK	107K12					107K10	107K12
2122-724	DREAM LODGE	106L10	99W08	105W08	105W08	104K08	103L10	106L10
1-1	PHILATELIST	103L12					103L12	101L10
211-	ZAHAM	103K08				97L08	103L08	103K08
0-0	CHARLIE TOKYO	100K11					100K11	94W08
1-	RED MOLONEY	95D10						95D10

I had been sweet on the chances of Gerard Butler's filly Silver Pivotal ever since she'd won a handicap over this course and distance on her return from eight months off in January, and I had also been all over her like a dirty little rash in my King Of The Sand column the previous Tuesday. What I really liked about that previous performance was that she had put in a sustained effort tight against the inside rail at a time when racing on that part of the track appeared to be a huge disadvantage. In my book, because of that she could be rated even better than the pure form and I wasn't going to desert her now, despite her wide draw.

Having listened to her rider Jamie Spencer discussing the race on Channel 4 Racing earlier in the afternoon, I got the distinct impression that he was very confident and that he wasn't unduly concerned by the outside stall 14, whereas other people were. Besides, he had overcome the same draw in order to win this race aboard Sri Diamond in 2006, so he knew what he was doing. I thought that 3-1 was a cracking price given all the positives and one (in my opinion) small negative and felt that she should have been much shorter. The draw was obviously on many people's minds, but not on mine so I snapped the price up.

Whether having to be tucked in from her wide draw was the difference between victory and defeat is impossible to say, but she came from a long way back with a devastating late run down the wide outside, only to find that Hattan was already home and hosed. I had told myself that I wouldn't blame the draw if she were beaten, though interestingly after the race her rider was inclined to do so.

Hattan holds off Silver Pivotal to win the 2008 Winter Derby

The next race on the card was a five-furlong Listed race. The favourite Excusez Moi had broken the course record when winning over six furlongs here last time, but although I respected him I wasn't sure that the shorter trip was ideal and that he was short enough in the market. I decided to abstain from the contest and was glad I had done so when Excusez Moi completely fluffed the start before running on to finish a never-nearer fourth.

The sixth race on the card was a nice one-mile-four-furlong handicap and I couldn't see beyond two of the horses taking part.

4-20 CORTAFLEX HANDICAP (0-95) (3) 4YO+								
12 furlongs		**Par = 107**			**8 runners**			
AW FORM	**NAME**	**BEST**	**LAST SIX**					
301-2334	SGT SCHULTZ	113L12	94K12	93L12	109L12	108L10	113L12	98L10
1	JOHN TERRY	113L12						113L12
0122-262	POLISH POWER	111L12	101L12	111L12	96L12	103L12	109L12	107W12
1/0111-3	GRANDE CAIMAN	110L13	98L08	95K08	103L13	110L13	96L12	103W12
37-35642	FUSILI	110L10	100K10	100L10	92L10	108L10	109L10	108L12
9064-39	MURFREESBORO	106L06	100D06	97K06	106L06	94K07	95W08	95W08
113-3315	TILAPIA	105K08	103S08	05K08	100S07	101S08	100K11	102L12
0-4	REGIONAL COUNSEL	103L10					95D07	103L10

John Terry (whom I had tipped for this race in King Of The Sand earlier in the week) had won a similar event over course and distance the previous month, a race in which Sgt Schultz had finished an unlucky third. This was an example of when split times were of value to me also, as that race had a pace rating of –7, which told me the contest was truly run (something of a rarity in races over the longer distances at Lingfield) and therefore the form was reliable. I felt there was every chance that Sgt Schultz could gain his revenge this time, but after only one outing on sand there was also a chance that John Terry could improve on that initial effort, so I had to back them both. I managed to get 5-1 about Sgt Schultz and 9-2 on John Terry.

John Terry travelled really well up with the pace and gave himself every chance, but Sgt Schultz got a clear run this time and made it count with a narrow if fairly comfortable victory. We were now off and running!

The penultimate race on the card was a maiden over a mile.

4-55 CEMPLAS MAIDEN STAKES (5) 3YO+

8 furlongs — **Par = 103** — **12 runners**

AW FORM	NAME	BEST	LAST SIX					
3	ATHEER DUBAI	102L07						102L07
3	BARTERCARD	97L07						97L07
56	CANARY ISLANDS	92W08					91L08	92W08
774-	MY SHADOW	90L08				78W07	81L08	90L08
3	ALWAYS A ROCK	87W07						87W07
6	SWEET DEMERARA	83L12						83L12
0	CAMERA SHY	63K07						63K07
	LA ZARZA							
	GREEN WADI							
	SUNDOWNER							
	YES EIGHTEEN							
	LAURENTIAN LAD							

Perhaps I shouldn't have got involved with so many unexposed types in opposition, but I was taken by the speed figure earned by Atheer Dubai in his previous try here and the stable had just won the Winter Derby, so at least they were in form. I backed him at 6-1 and that looked a good price when he was then backed into 7-2, but it didn't look such a big price when after holding every chance he could only manage a one-paced fifth behind the favourite Sundowner, who was representing the same trainer/jockey combination as Silver Pivotal.

5-30 LINGFIELD PARK FOR WEDDINGS HANDICAP (0-85) (4) 4YO+

8 furlongs — **Par = 105** — **11 runners**

AW FORM	NAME	BEST	LAST SIX					
260263-5	DICHOH	107S08	98K08	98L08	102W08	101K08	100L08	98S07
-9362235	DUDLEY DOCKER	107S07	104W07	100L08	103S08	104S07	101S07	95W07
1776-131	HIGHLAND HARVEST	106W08	93W08	89L10	97L08	104L08	106W08	97L08
-3321111	NINTH HOUSE	105W08	99W09	103L08	103L07	93W08	100W08	102W07
454-1552	BEE STINGER	105L08	95W08	100L08	101L08	94L08	97L10	96K10
11242-62	LAWYERS CHOICE	105K08	105K08	100K08	100L10	96L08	97K08	97L07
3/2-77	APACHE DAWN	104K08	100W07	104K08	97L08	91K07		

3141617-	HESSIAN	103L07	99L08	98K08	103L07	100L08	94W07	92W07
93/8-	TRIMLESTOWN	98L07	80W07	93L07	98L07			
3/8-4	ZAFONICAL STORM	97L07	71L05	97L07	94L08			
	TWILIGHT STAR							

The final race of the day was a nice-looking handicap, again over a mile.

I was very sweet on the chances of Highland Harvest who had been successful in both of his outings over course and distance this year, so I was more than happy to take 4-1 about him. He looked all over the winner soon after turning into the straight too, but for some reason he appeared to hang fire and was run right out of it, eventually only finishing fourth behind 20-1 shot Dudley Docker.

This wasn't quite the start to the journal that I had been hoping for, though the margins between success and failure hadn't been that great.

Friday 28 March – Kempton

I have fond memories of Kempton Park going right back to my early childhood. I was there when the first ever lady jockeys' race took place at the track in 1972 and can remember the horses racing wide (very wide) around the home bend before the late Meriel Tufnell steered Scorched Earth to victory and the combination made history. I can also remember when Charisma Records started to sponsor the October jumps fixture a couple of years later and a pop concert used to follow the racing. The place used to be crawling with hippies, though as I recall it the bands that played there weren't exactly household names. I wonder whatever happened to Chilly Willy & The Red Hot Peppers?

When Kempton's grandstand and other buildings were refurbished in the late 1990s, there was a period when I wasn't too keen on the place as I thought it had lost some of its character, but I love it again now especially since the Polytrack was laid.

One claim to fame that Kempton certainly has is that the toilets have the most powerful hand-dryers in the developed world, a fact that I tend to forget until the very last minute and today was going to be no exception. Whilst availing myself of that particular facility, the downward pressure on my palms was so great that my arms were almost wrenched from their sockets.

Emerging from the gentlemen's lavatory looking like a gibbon, I tried to decide on how I was going to play the first race of the evening, an ordinary looking five-furlong handicap.

6-20 INTERCASINO.CO.UK HANDICAP (0-75) (5) 4YO+

5 furlongs		**Par = 103**				**7 runners**		
AW FORM	**NAME**	**BEST**	**LAST SIX**					
4612012-	SPIRIT OF CONISTON	108K05	94S05	100W05	100W05	81W05	108K05	96W05
729-2637	DIMINUTO	105L05	100W05	89S06	105S05	93W05	104S05	105L05
2101-458	WHAT DO YOU KNOW	105K05	99K05	91W06	105K05	94L06	97L05	104L05
0/5/5-	HOLBECK GHYLL	103L05				56W07	96L05	103L05
61119-18	FAST FREDDIE	101W06	101W06	98W05	96L06	87S06	96K05	97W05
0/67/7-	DIGITAL	100L06			100L07	100L06	99W06	99K07
49/5-9	GLEAMING SPIRIT	93K06			90W05	82W06	92S06	93K06

This wasn't the easiest sprint handicap to dissect. Both Spirit Of Coniston and What Do You Know had winning form over course and distance and both had managed to win from unfavourably low draws, which is a difficult thing to do, but it was hard for me not to use my own draw statistics as my main tool here.

As has been shown earlier in the book, the five-furlong course at Kempton has one of the biggest draw biases at any All-Weather track. Coming into this race the rails draw boasted a 37% strike rate in terms of winners to runners in the previous 12 months, so Fast Freddie just had to be supported on that basis alone. I also hoped that he would try and lead early, something he had done to maximum effect before including on his last visit here. Under the circumstances I thought that 4-1 was a massive price and was happy to take it, though he did drift out another half-point before the off.

Unfortunately he wasn't able to lead this time though he did race handily, but that isn't quite the same thing and he never really looked like winning. The race eventually went to Holbeck Ghyll from the fast-finishing Digital.

The second race was quite a competitive looking handicap over a mile and two furlongs.

6-50 PLAY BLACKJACK AT INTERCASINO.CO.UK HANDICAP (0-75) (5) 4YO+								
10 furlongs		**Par = 103**			**10 runners**			
AW FORM	**NAME**	**BEST**		**LAST SIX**				
325/21-0	SARWIN	108W09	102K10	101W12	94K10	08W09	106W09	97K10
-1321224	GIVEN A CHOICE	108L12	102W09	101W09	96K11	102K12	91K08	101W09
/6127-01	DINNER DATE	108K10	96K10	102K10	100K10	90L10	84K10	100K08
3/63-472	LORD THEO	107K10	95W06	101L07	99W08	99L08	67W08	107K10
-1323425	STEIG	106W12	96K11	106W12	99W09	101W08	102W09	84S08
18/007-1	STEELY DAN	106L07	104L08	97K10	99L07	91L08	85L08	103K11
2262-153	CHIA	104W08	103W09	96K08	97W09	68W09	98W09	88W09
4432-21	EMPEROR COURT	103L10	95L08	93K10	96W09	103L10	100L10	99L10
10846/9-	TRANS SONIC	101L12	97K08	90K08	93K07	95K06	101L12	95L08
	HAARTH SOVEREIGN							

Given A Choice was consistent and had plenty of form around here. I was tempted by him, but preferred Lord Theo who had recorded such a decent speed figure when runner-up over course and distance the previous month. If he reproduced that he would win, and I was more than happy to take 8-1 about that eventuality.

He nearly did win too, but after holding every chance against the inside rail he was just beaten into third by sand-debutant Haarth Sovereign and the favourite Emperor Court.

The third race on the card was a one-mile claimer.

7-20 INTERCASINO.CO.UK CLAIMING STAKES (6) 4YO+								
8 furlongs		**Par = 101**			**9 runners**			
AW FORM	**NAME**	**BEST**		**LAST SIX**				
93852-56	BARRY ISLAND	106L10	102K11	96L10	97L10	85L10	97L10	105L10
1/70/17-	CACTUS KING	104L10	96L08	102L08	100W09	99W09	104L10	83L10
3/2-770	APACHE DAWN	104K08		100W07	104K08	97L08	91K07	95L08
07-161	DUSHSTORM	103L08		94W09	96K08	103L08	94K08	100L08
5828-375	SHOULDNTBETHERE	101L10	94W09	99K08	87S08	94K08	88K08	88K08
257-8113	NOT NOW LEWIS	101K08	90W09	94W08	70W08	101K08	100K08	99K08

7/430-5	BED FELLOW	100K08		80K07	100K08	100K08	91K08	13K12
2242/6-8	VICTOR TRUMPER	96S08	87L08	96W07	93W07	96S08	84L10	66W08
84290	FILM QUEEN	89L07		85K07	76L07	85L06	89L07	85L07

I don't normally like claimers. The horses that contest them are either there because their connections are keen to get rid of them or they are considered to be badly handicapped. Either way, these races always look dodgy betting propositions to me and the only reason I got involved this time was because I considered top-rated Barry Island to be far too big at around 14-1. Richard Quinn was taking the ride and I was certain that he had a much better chance than his odds might suggest, even though he was the old man of the party.

The only concern I had was that he was stepping down in trip, so he would need a strongly run race in order to compensate for that. So imagine my horror when, after the runners had exited from the stalls, the leaders set one of the slowest early tempos I had seen in many a year. The sectional rating I eventually allocated this race was +76!

Barry Island never had a prayer, but at least I knew my fate early. The race was won by joint second-top-rated Cactus King at 9-1 from the other joint second-top-rated Apache Dawn at 11-4. In retrospect perhaps I would have been better off combining the top three in exactas.

The fourth race was a maiden for three-year-old fillies. There wasn't enough form to go on so I went for a drink and had a longer and closer look at the fifth race on the card, a seven-furlong handicap.

8-20 BIG JACKPOTS AT INTERCASINO.CO.UK HANDICAP (0-60) (6) 3YO

7 furlongs		Par = 96				12 runners		
AW FORM	NAME	BEST	LAST SIX					
5-686723	KAMAL	96W05	88K07	82K07	87S06	88W05	96W05	94K06
58496-48	LADY SANDICLIFFE	96L10	82W07	88K07	86K10	91K08	87L08	96L10
40-54625	TOO GRAND	96L06	81K07	90L07	90K08	87K07	77L07	92K06
416	SNOW BOUNTY	95L08				76L07	82L08	95L08
443-5223	LLAB NALA	94L07	88K08	82L08	70K06	82L08	91L07	89L07
7739	MARQUIS DE LOUVOIS	91K06			91K06	78S06	88S06	82K06
7605	MISS TILEN	89S08			84W07	89S08	71S08	88W09
635655-7	OCEANA BLUE	88W07	87W08	87K08	85L07	88L08	88W07	86K07
7-0488	HERRBEE	88K08		80L06	82K07	88K08	79K08	87L07
690-	BAHIA PALACE	87W06				87W05	87W06	77W07
85-9443	SHABNAAM	86W06	81W06	81W06	86W06	80L06	78S08	84K07
	FOLLOW YOUR SPIRIT							

Seven furlongs is another trip at Kempton where horses drawn against the inside rail have a decent record. On this occasion it was the favourite Kamal who had the plum draw and he was being backed, so even though 5-2 was short enough for a race like this it was just about tempting enough, so I waded in.

He was backed in still further to 2-1 at the off, but even though he was always in a good position and held every chance, the eventual winner Oceana Blue was running all over her rivals from a very long way out. Kamal couldn't even hold off Too Grand for second.

The penultimate contest was a two-mile handicap and this was a race I really wanted to get stuck into.

8-50 PLAY ROULETTE AT INTERCASINO.CO.UK HANDICAP (0-70) (5) 4YO+

16 furlongs | **Par = 103** | **12 runners**

AW FORM	NAME	BEST	LAST SIX					
28311-84	FEATHERLIGHT	108L12	98K11	100L12	102K16	103K16	73L13	96K16
56527-46	BEST SELECTION	106K11	101K11	96K12	100K10	102L10	87K11	96K12
32/47-13	CODA AGENCY	105S12	95K16	93L16	100K16	98L16	105S12	99S12
74257-67	ARABIAN SUN	105L16	98K16	104W16	101K16	102W16	105L16	88W16
465899-4	LYSANDER'S QUEST	105L16	97W16	91K16	98K16	99K16	90K16	105L16
3339065-	FOLLOW ON	105K16	100L13	99K16	86K12	95L16	99K16	101K16
15-66322	LORIKEET	104K12	87L10	96L10	96L13	104K12	101L16	97K16
5135-163	MISTER COMPLETELY	103W13	103W16	103W13	88S16	98W16	96L16	96K16
6/0/	BRIGADORE	100L05					100L05	96K06
02/0-	KICK AND PRANCE	99W12				80K08	99W12	71W13
/9280-00	LAY THE CASH	98L10	75L06	97L06	94L06	83K06	98L10	87K12
69-	PRINCESS AIMEE	93W12					86W09	93W12

Featherlight had won three times here – two of them over this trip – and had never scored anywhere else. She had shown signs of a return to form in her last start and looked a tempting wager at 6-1. She had looked a potentially nice bet when I had tipped her in King Of The Sand at the five-day entry stage, so I wasn't about to desert her now.

I wanted to back Mister Completely too, as he had also won a couple of races over this course and distance and was available at 8-1. I also combined the pair in reverse exactas.

I certainly couldn't complain about the ride Featherlight was given. She was produced with her effort at just the right time, but wasn't good enough to cope with Coda Agency and had to settle for second. Mister Completely would have finished much closer than sixth had he not been squeezed out when trying for a run up the inside rail halfway up the home straight, so this was definitely a case of nearly but not quite.

Until the end of 2006, the last race on the card would have been known as a banded stakes and however you may flower it up by renaming it as a classified stakes, it's still crap.

9-20 INTERCASINO.CO.UK CLASSIFIED STAKES (0-45) (7) 4YO+

7 furlongs | **Par = 99** | **13 runners**

AW FORM	NAME	BEST	LAST SIX					
86-54728	A TEEN	103K07	88L07	85S07	93K07	93K07	93K07	92K07
-3535634	ONLY IF I LAUGH	102W06	92K07	94K08	92W07	94K07	92K07	97K07
079-8821	STRAIGHT FACE	101K07	85W07	88L08	87K08	84K10	97K08	101K07
46/936-2	PRINCE VALENTINE	99K07	91K10	93K10	89L07	94L08	95L07	99K07
06/6-694	CYFRWYS	98W06	82S06	94W06	96W06	93W06	83W07	91W07
0485-7	RAISE AGAIN	98L10		89K08	92K07	98L10	91S08	93K07
-0672252	SUHAYL STAR	97W06	89K07	94K07	95K06	95W07	92K07	97W06
0600-087	PAJADA	97L07	97L07	89L07	87L10	89K08	93K07	86K08
5-036549	DOCTOR NED	95L08	81K06	86K07	95L08	85K06	92K07	88K07
0-56P566	MTOTO GIRL	94K11	90K08	94K11	0K11	93K08	86K10	92K07
9/04/	CANARY GIRL	90W06				79S06	46W08	90W06
7603	NORTHSTAR EXPRESS	87K08			85W07	87W09	87K08	85K08
	TVARA							

Still, as many hard-working folk will tell you there can be money in crap and I was very hopeful that would be the case here.

Only four of these could win according to my figures and one of those, A Teen, could be safely discounted as he had finished a long way behind Straight Face here last time and there was no obvious reason why he should get any closer. Straight Face's chance was obvious looking at the ratings, as he was relatively progressive and a reproduction of that last figure would be enough to win this.

Prince Valentine was also a player, but whereas he had earned a rating from me last time that was good enough to win a race like this, Only If I Laugh had exceeded the class par on a few occasions so I just preferred him. I backed Straight Face at 5-2 and Only If I Laugh at 9-1 and combined them in a reverse exacta.

Although Straight Face was always in a good position, it looked for a long time as though the rider aboard Suhayl Star only had to press the button in order to win such was the ease with which he was travelling, but when the gun was finally put to the horse's head he found absolutely nothing. Straight Face kept on pulling out a bit more and just managed to hold off Only If I Laugh and Prince Valentine who passed the post locked together just a neck behind the winner. To my great delight, Only If I Laugh just got the better of the photo for second so the evening had ended on a high.

Friday 4 April – Lingfield

Looking through this card, it wasn't the most competitive Lingfield Park had ever staged, but that had been the case with quite a few recent All-Weather fixtures. Some people had argued that the reason for the small fields was because there were too many All-Weather meetings taking place and the product was therefore being spread too thinly. It was hard to disagree with that as even though in theory we were at the start of the turf Flat season, only a sprinkling of meetings had taken place on grass since the Doncaster Lincoln Meeting on March 22nd and on many days there had been none at all. For instance, the previous Friday (if you included Dundalk) there were four meetings on artificial surfaces taking place in the British Isles, plus just one jumps fixture.

Today there were three All-Weather meetings (again including Dundalk) plus a turf Flat fixture at Southwell, though even there two of the seven races were taking place on Fibresand. The only other meeting was the second day of the Aintree Grand National meeting. Whose fault was this? Opinions on that varied, mainly depending on whether you were a fan of All-Weather racing or not, but if the reason was that the turf tracks were reluctant to race at this time of year for fear of ruining their ground for the season to come, then the All-Weather tracks were merely filling a vacuum.

This meeting had been transferred from the still not yet open Great Leighs and, as with most fixtures that don't appear on the calendar, the crowd was pitifully small. In fact the number of people there was probably about enough to throw a cordon around a telephone box, though that wasn't necessarily a bad thing on a warm sunny day like this.

The first two races on the menu were both maiden fillies' events, one for older horses over ten furlongs and the other for juveniles over five furlongs. The first race didn't particularly turn me on, but I thought that paddock inspection might be a valuable guide for the two-year-old event.

2-45 EUROPEAN BREEDERS' FUND MAIDEN FILLIES' STAKES (5) 2YO			
5 furlongs		**Par = 89**	**6 runners**
AW FORM	**NAME**	**BEST**	**LAST SIX**
4	DR WINTRINGHAM	82K05	82K05
	RAGGLE TAGGLE		
	KATE THE GREAT		
	ALPHABETH		
	MAYBE BLUE		
	AMOSITE		

The value of being able to watch horses in the parade ring isn't confined to All-Weather racing so there are no magical tips here that you won't find in many other publications. Whatever type of racing you like, if you get the chance to see the runners in the flesh beforehand, especially in races confined to inexperienced horses, the advantage you will hold can be invaluable.

I am no great paddock judge, but I can certainly tell a ready horse from one that is backward or going to need the run and/or the experience. On this occasion one of the fillies, namely Kate The Great, stood head and shoulders above the rest despite being one of the three newcomers. She was nicely on her toes and looked on very good terms with herself in the paddock, but she was also being very well backed in the ring and 3-1 was the best price I could get.

Three of her rivals had run before though only one, Dr Wintringham, had done so on sand, so just in case Kate The Great needed it more than I thought I also backed her at 5-2.

Happily my judgement of Kate The Great proved right, though only by a nose from fellow-newcomer Raggle Taggle, with Dr Wintringham back in third.

The third race was yet another contest confined to fillies, this time a six-furlong handicap.

3-20 BACK HERE TOMORROW FILLIES' HANDICAP (0-95) (5) 3YO+								
6 furlongs		**Par = 100**			**6 runners**			
AW FORM	**NAME**	**BEST**	**LAST SIX**					
6/90-667	LINDA GREEN	103K06	101K06	95K06	93W06	103K06	84S06	97K06
2-482203	MAGGIE KATE	101L06	86L06	91K07	91K07	101L06	88K07	94L06
3321/12-	FEELIN FOXY	98W05	90L05	89L05	94L06	91K05	98W05	98W05
410-	PENRICE CASTLE	94L05				87K06	94L05	81K06
5-	CINNAMON HILL	93L07						93L07
046-	FOREVER CHANGES	85L06				73K05	85L06	83L05

I had been very keen on this race earlier in the week, as I believed a filly called Tarkamara had an outstanding chance. I had napped her in King Of The Sand and was understandably delighted when I saw that she had been declared to run at the 48-hour entry stage. I was all prepared to have a decent bet on her, but alas she was then declared a non-runner so I had to take another look at the contest.

I didn't like either of the fillies that eventually went off joint-favourites. Penrice Castle had it to do on my figures, whilst the reason I was against Feelin Foxy was rather different. As I have said before I don't tend to use weight as a major factor in my calculations, not so much handicap marks which have more to do with class than actual weight carried, but all this business with Horse A being 7lb better off with Horse B for a two-length beating, etc. I do, however, get nervous about Flat horses carrying more than ten stone outside of amateur-rider races, which I tend to swerve in any case. Not many of them seem to win, and as Feelin Foxy was carrying a 6lb penalty for a recent win on turf taking her weight up to 10st 4lb, I was happy to take her on.

The pair I focussed on were Linda Green and Maggie Kate, who as my ratings show were the only ones to have previously exceeded the class par on sand and were currently trading at 6-1 and 7-1 respectively. There was evidence that Linda Green was better suited to Kempton, but at the available odds I was willing to give her the benefit of the doubt.

I was right about the two joint-favourites, but whilst my selections both gave themselves every chance, Cinnamon Hill got in the way and held off Linda Green by a neck with Maggie Kate back in fourth. Looking at the race subsequently, especially after I had allocated Cinnamon Hill a speed figure of just 91 for her win, I couldn't help thinking that Tarkamara would have walked it.

The next race on the card was a Class 2 handicap over two miles, but unfortunately only three horses went to post and I was undecided as to whether I should get involved or not.

3-55 LINGFIELD PARK FOR WEDDINGS HANDICAP (0-100) (2) 4YO+

16 furlongs | **Par = 109** | **3 runners**

AW FORM	NAME	BEST	LAST SIX					
124241-2	SALUTE	112L12	107K16	108K16	106W13	105W13	103W13	109W12
340/34-	NODDIES WAY	105L16		100L13	99L16	0K16	105L16	99L16
3-317111	ROLLIN 'N TUMBLIN	104L16	101L10	103K11	90K11	86L13	104L16	98K16

Rollin 'N Tumblin was trading at around 1-2, but on my figures he had it to do class-wise so he would have been a decent lay for those so inclined. The problem was that I could visualise the race turning into a messy affair and that the favourite could have the contest fall into his lap if it turned into a sprint. My eye kept on being drawn to Salute who, on sand at least, had on eight occasions earned a speed figure higher than anything his two rivals could muster, but he was going to need a proper gallop.

The outsider Noddies Way couldn't possibly win, could he? It's true that John Panvert's gelding had become something of a minor celebrity having made his racecourse debut in the 2006 Derby and then gone on to run 18 more times (including once over hurdles) without managing to get his head in front. Even with the possibility of a false pace, surely he couldn't get off the mark in this race.

In the end I weakened and went in on Salute at 3-1, but once the race was under way I was very pleased to see Noddies Way set a decent tempo considering the circumstances, and the bonus was that despite the solid gallop the favourite was taking a keen hold out the back. It seemed as though Salute could pick the leader up when he wanted, but then almost unbelievably just as Salute was brought out to challenge, Noddies Way found a bit more out in front and kept on all the way to the line to hold off Rollin 'N Tumblin, who just got the better of Salute for second.

I was shell-shocked. Noddies Way had broken his duck at the 20th attempt in a £16,000 handicap.

The next race was the main event on the card, but I didn't like the contest as a betting opportunity. Only a couple of the runners had been to Lingfield before and their form wasn't great. Those near the head of the market had shown their best form at Kempton, including Mr Aviator who had finished second in the Rosebery Handicap just six days earlier. He looked the obvious one, but he never went odds-against so I decided to sit the race out. The favourite duly won comfortably, but I was still just happy to watch him do so.

Still reeling from the victory of Noddies Way, I decided to go and have a drink and sit outside whilst taking another look at the next race, a low-grade but quite competitive handicap over a mile and five furlongs.

5-05 LINGFIELD PARK FOR CONFERENCES HANDICAP (0-65) (6) 4YO+

13 furlongs — **Par = 101** — **12 runners**

AW FORM	NAME	BEST	LAST SIX					
5119166-	ADAGE	107W13	103W13	107W13	96K16	101W13	101W13	97L16
6527-463	BEST SELECTION	106K11	96K12	100K10	102L10	87K11	96K12	100K16
/8440-83	MONTOSARI	105L16	98L12	82K12	102K16	0L13	69K16	92L13
/212350-	BOB'S YOUR UNCLE	105L12	100L12	105L12	104K16	103L13	99L16	86K16
3-243112	CAPITALISE	104W16	97W16	99K16	100W16	100L13	95L13	100L12
/750/30/	GREAT VIEW	103W07	95L13	89L13	93W13	86L10	98K12	84L12
430/769-	BOBSLEIGH	103K16	102K16	103K16	94L16	98K16	96K16	96L13
820-8	CAVALLINI	103K11			96K16	101K12	103K11	98L10
347622-6	PARTY PALACE	101W13	90K11	87K11	97K12	101W13	95W13	98K16
0/07/473	BARNBROOK EMPIRE	100L10	85L13	88W13	84K12	91L13	91L13	99L13
5542-610	WHAXAAR	98K12	84K08	91K12	98K12	90K12	95L12	92K12
568	SPELLMAN	90L10				77S11	90L10	82W12

Sitting there with my cold drink on a very warm day, I felt that all was well with the world and I was just thinking how idyllic things were when the peace was shattered by the unmistakable (and very noisy) sound of a thunderous fart. Not only was the atmosphere shattered, my table actually shook with its force and my radar was immediately drawn to the very elderly gentleman who had his back to me on the next table. Guilty he certainly was, but he maintained his composure whilst the people on the other tables tried to do the same. I looked around at each of them, daring them to make eye contact with me so that they might break down, but they never did. They remained totally focused on their racecards and newspapers, though those items were drawn much closer to their faces than they would otherwise have been.

Just like being dumped on from high above by a bird, perhaps this was a sign that the elderly gentleman was going to be lucky, certainly luckier than those of us downwind of him, but I quickly had to gather myself together as the next race was closing in fast.

The two I liked were Bob's Your Uncle and Capitalise who had both shown a liking for middle distances at this venue. Bob's Your Uncle was returning from five months off so there was a slight question mark against his fitness, but he looked fine to me in the paddock and I was happy to stick with him. I backed Capitalise at 9-4 and Bob's Your Uncle at 11-2.

Both ran well with Bob's Your Uncle finishing third and Capitalise fourth, but the Gary Moore-trained Cavallini was the subject of a gamble and they don't normally get it wrong when the money is down. Backed in from 7-1 to 100-30, they didn't get it wrong this time either.

The last race of the day was a handicap over a mile with just the five runners.

5-40 ARENALEISUREPLC.COM HANDICAP (0-70) (5) 4YO+

8 furlongs **Par = 103** **5 runners**

AW FORM	NAME	BEST	LAST SIX					
1556-676	SIR LIAM	105L10	104L10	93K10	98K11	90K11	105L10	90L12
1/96/60-	DANIEL THOMAS	105L08		96L07	105L08	101L08	102L10	93K08
71-13102	JOY AND PAIN	105L07	98L07	95L07	105L07	98L07	90K08	99L07
0/3-82	SCARLET FLYER	100L08			87K08	100L08	98L07	100L08
52/2	TINNARINKA	98L10				89K07	87K07	98L10

The first two in the market were the bottom two on my ratings, which meant that I had to take them on, but with whom?

Joy And Pain had won seven times on the Polytrack here, but never at beyond seven furlongs. However, I could see this race being run at a crawl thanks to the small field and that would enable him to use his finishing kick, so I decided to support him at the available 4-1.

My feeling that they would go a dawdle was quickly shattered when Sir Liam, who was dropping in trip, was given a very positive ride and made it quite a fair test. In truth Joy And Pain was beaten before stamina became an issue and he trailed in a well-beaten fourth behind Tinnarinka.

This had not been a great day punting-wise, but I wouldn't have to wait very long for a chance to get back on the straight and narrow.

Saturday 5 April – Kempton

Today was Grand National day. I had wanted to watch the big race in comfort, but I also wanted to go to an All-Weather meeting today. Lingfield were staging a nice-looking fixture in the afternoon and I could have gone there and watched the National on one of the TV screens, but I decided to both have my cake and eat it.

If I played my cards right, there was just enough time to watch the National in my own living room and then jump into the car and drive down to Kempton for their evening card. Fortunately the big race went off more or less to time and after having watched Comply Or Die's wonderful triumph, I eventually arrived at the Sunbury track with at least half an hour to spare.

The first two races on the card were both claimers, not normally my sort of race, but whilst I was happy to swerve the first of them I did have an angle on the second.

6-50 MICHAEL SMALL 21ST BIRTHDAY CLAIMING STAKES (6) 3YO+

6 furlongs **Par = 101** **4 runners**

AW FORM	NAME	BEST	LAST SIX					
1-215238	ANGEL VOICES	104K07	100K07	104K07	96K07	96L06	97L06	95L06
/0-34223	RED RUDY	104K06	98W08	104K06	103K06	104K06	103K06	88W08
-2747535	GOODBYE CASH	103L06		97K06	83S06	87S07	96W06	83W06
36-	HONEY MONSTER	91L06					91L06	86W06

Whilst there wasn't much to split the top two on my ratings, there was no doubt in my mind that Angel Voices was going to get the early lead, and a pretty soft one at that if the four horses' running styles were analysed. Angel Voices had gained both of her wins at this track, one of them over this trip, and on each occasion she had made every yard under today's rider Declan Cannon. For that reason I was prepared to forgive her modest effort at Lingfield in her previous start.

She looked pretty bombproof to me so although 6-4 is below the sort of price I normally like to bet at, it did look value considering how the race was likely to pan out.

Things did pan out perfectly too. Angel Voices bounced straight into the lead, travelled very nicely out in front, and still looked to have plenty in the tank with an advantage of a couple of lengths crossing the intersection where the inner and outer loops meet. The real bonus was that Red Rudy, whom I considered to be the only real danger, fluffed the start and then took a fierce hold out the back.

It occurred to me that only an act of God could stop Angel Voices from winning, but God moves in a mysterious way and he must have had the right hump with Angel Voices because she stopped to nothing very quickly, allowing her three rivals to sweep past her in the last furlong, headed by Red Rudy.

I was genuinely surprised and still can't work out what happened. Mares do have off-days and this was certainly one for Angel Voices.

The third race was a 12-furlong all-aged maiden and I'm not a great fan of these races as a betting medium, so I just watched with no financial involvement.

The next race was a handicap, again over a mile and a half, but this time I was interested.

7-50 DIGIBET CASINO HANDICAP (0-55) (6) 4YO+

12 furlongs **Par = 101** **11 runners**

AW FORM	NAME	BEST	LAST SIX					
0636/077	SHAMI	108L10	93S11	95S08	89K10	87S11	85W08	92S08
750-9336	AMWELL BRAVE	107S12	100K12	91K12	89K12	102K12	99S12	97W12
515146-0	SUMMER BOUNTY	102L10	87W12	99K10	86K11	96K11	96K12	96K12
/28/7213	BEN BACCHUS	101K12	89S12	91K16	98K12	94L12	101K12	94S12
70-837	NANOSECOND	101K12		77W07	95L10	82K16	101K12	91W09
1144	LOUGH BEG	100W09			92K10	98K10	99K10	100W09
/17-1254	SAHF LONDON	99K12	96K11	90K10	98K10	97K10	91K11	99K12
5/75-326	NIQAAB	98W12	86L07	93W09	88K12	90S12	98W12	89L12
9280-000	LAY THE CASH	98L10	97L06	94L06	83K06	98L10	87K12	90K16
037404-1	TIEGS	96S12	91L13	86S12	85S14	71S12	95L12	95K12
0	LAURENTIAN LAD	79L08						79L08

Like Angel Voices, Lough Beg had gained his two career victories at this track which certainly made him a player, but I had a niggling doubt over him as he had never attempted this trip before on the Flat and with a decent-sized field like this, I felt there was a good chance they would go a solid pace. There were no such doubts over Amwell Brave however, even though a record of just three wins from 75 starts was hardly going to set the pulse racing. Odds of 13-2 about him looked fair enough to me though, so I got stuck in.

As things turned out, perhaps it was easy to spot why Amwell Brave had such a poor win record as he only does things very slowly. He was held up early and was asked for his effort soon enough, but the headway he was making was only very laboured and William Buick did extremely well to force him into third place behind Lough Beg, who appeared to get every yard of the trip. The sectional rating I gave the race was +41 though, so they hadn't gone that fast early after all.

The fifth race on the card will live with me for longer than it should and I doubt that I will ever be allowed to forget it.

8-20 DIGIBET SPORTS BETTING HANDICAP (0-52) (6) 4YO+

8 furlongs | **Par = 101** | **14 runners**

AW FORM	NAME	BEST	LAST SIX					
50/0/076	KING OF DIAMONDS	107L12	99L08	101L12	99L08	88K12	94K10	99K12
4-226348	JOMUS	107K10	97K08	92K08	99K08	102K08	105L10	96K10
5-326759	MISTER BENJI	106S08	93S07	96W08	97S08	93W07	86W08	96W08
09/802-0	BOLD PHOENIX	104S08	76K10	89W09	88S12	87W09	97W08	96L08
57/3-	FOREIGN EDITION	103W06				103W06	99W06	94K07
607/00-4	DA BOOKIE	103L08	98L08	91K08	85W09	94L08	94W08	98L08
7-453675	POSTMASTER	102L08	94K08	91K08	98K08	100K08	96L08	88K08
0/9-9502	GRENANE	100W08	87W06	71W07	94W07	88W08	78W08	100W08
52068-87	BODDEN BAY	100W06	95W06	66W07	88S06	95K06	75S06	91W07
902-006	KORTY	99K08	90L07	70K08	99K08	78K08	83W09	97L08
40/09/	SHROPSHIRELASS			92L10	91W05	47S07	78W06	92L10
5000-	THE GRAIG			92L10	77K11	65S11	92L10	86K12
08/8-	MAJOR LEAGUE				91W08	88K07	86K08	91W08
8/	EDGEFOUR					76L07	76L07	

We British love a glorious failure it seems, although such a description seems harsh in many cases. What we like to see is someone who tries their socks off whether they are successful or not, and the more unsuccessful they are the higher their celebrity status gets.

Noddies Way was becoming such a phenomenon before his win the previous day even though he hadn't been around that long and hadn't had that many tries in the overall scheme of things. The irony is that after Friday's victory the fascination surrounding him would have subsided, which it would also have done if the likes of Amrullah or Quixall Crossett had ever won a race, or if Eddie The Eagle had ever won a medal.

Jomus had at least won a race, but the trouble is that it was only one from 63 starts, 46 of them on sand. Because of that, and because he always seemed to grab defeat from the jaws of victory, he had made a bit of a name for himself not far below the prestige of the three heroes mentioned in the previous paragraph. Just to give an idea about how the gelding was viewed, the following had been written about him in the *Raceform Notebook* in some of his recent outings –

> *"…. has made a bit of a name for himself for all the wrong reasons"*
> *"….not for the first time, consented to hit full stride all too late in the day"*
> *"….ran his usual sort of race and remains one to avoid for win purposes"*
> *"….always runs like this and it would be a big mistake to get sucked in by him"*

After comments like that and with the evidence of my own eyes, how could I possibly consider backing him even though he had a big chance on my ratings? Earlier in the evening I had haughtily informed my wife and my sister, who were with me, all about this horse and that from our vantage point on the lawn we should be able to hear the barking from Jomus as he dogged it in the last 100 yards (though to be fair I did suggest that he might be a knocking place-only bet). I did get the impression from the way they looked at me that they felt I might have just set myself up for a fall.

As far as my own punting on this race was concerned, I much preferred the chances of Grenane and Postmaster and happily took 5-1 and 7-1 about them respectively.

Grenane didn't get the clearest of runs on the inside of the track when threatening to get involved and neither did Postmaster, though he probably still had his chance had he been good enough. The problem was that he ran into Jomus when everything happened to go right for him and, looking at the way he won so tidily, you would never have believed that he was considered the biggest "twicer" on the planet by many.

Noddies Way and Jomus winning within 30 hours of each other? What next? I knew this diary was going to be a challenge, but I didn't know booby traps like these were going to be strategically placed in order to test my resolve.

The next race would test me further.

8-50 PANORAMIC BAR & RESTAURANT LOYALTY SCHEME HANDICAP (0-80) (4) 3YO

8 furlongs **Par = 100** **9 runners**

AW FORM	NAME	BEST	LAST SIX			
67-11	GALLIC CHARM	103L08	92L07	73L08	85K08	103L08
5631	CANARY ISLANDS	98L08	91L08	92W08	98L08	89L08
59-1	COMMANDER CAVE	93L07		84L07	89L07	93L07
1-	CROSSTAR	92L08				92L08
093-	BLACK OR RED	89W07		84W07	80K07	89W07
9-	ALWAYS BRAVE	78W08				78W08
	SUMMON UP THEBLOOD					
	TOTALLY FOCUSSED					
	MIZOOKA					

Gallic Charm looked outstanding judged on that last rating and the bonus was that she had also previously won here. I had expected her to be pretty short in the market, so was very surprised that she drifted right out to 3-1. She just had to be backed at that price, but whether someone knew the race was going to become a mess right from the start and that is why she proved friendless in the market, then they were to be proved absolutely right. She walked out of the stalls and then pulled so hard out the back that she proved a nightmare ride for Jamie Spencer. At one point I thought she was going to run off the track, but although she didn't quite do that her race was over almost as soon as it had begun. Eventually she could only manage a modest fifth behind the winner Commander Cave.

I don't possess the air of confidence of many other professional punters or racing journalists. I don't mean that I'm not confident – I just mean that I can't emit that air of invincibility that many can even when things get desperate. I go through losing runs, some very long losing runs, but my confidence comes from the fact that when they do eventually finish I tend to emerge from the other end having at least compensated for the losses.

The period of this betting journal had been tough up to now and if it was like that in order to test my resolve, what happened in the last race was the reason I still like playing this game.

9-20 KEMPTON FOR WEDDINGS HANDICAP (0-60) (6) 4YO+								
7 furlongs		**Par = 101**			**14 runners**			
AW FORM	**NAME**	**BEST**	**LAST SIX**					
753424-4	FRANKSALOT	102L06	99K08	93L06	98L06	101L06	97W07	90W08
8/0-767	TAMINO	102K06		93L06	96L07	93K06	102K06	83S07
3627-	TYRANNOSAURUS REX	102K06			95W07	92W07	102K06	93W07
926-2314	SOLICITUDE	101W07	98L07	91W07	100S07	84W08	97W07	94L07
7-164136	AGGBAG	100W07	100W07	94W08	87S07	98W07	88W08	97W07
/56165-6	WADNAGIN	100L07	84L06	93W06	97W07	89W07	93K07	100L07
5371270-	FORCED UPON US	100K07	95L08	92K08	100K07	97L06	97L07	74W06
8-221619	JESSICA WIGMO	100K07	92L08	98K08	93L07	89K07	100K07	88L07
71756399	BUZZIN'BOYZEE	99W07	88K07	89S07	70S06	97W07	92W07	86S07
44-	SINTENIS MAC	94W06					94W06	90S06
6-647	COLOSO	93L08			87K07	91K07	92K08	93L08
7	MAN OF LETTERS	91W09						91W09
842905	FILM QUEEN	89L07	85K07	76L07	85L06	89L07	85L07	80K08
3/9-	NAMIBIAN PINK	87K08					87K08	86W08

This was another modest handicap, but still quite competitive with the favourite trading at around 4-1. Having gone through the race a few times I had come down on two possible winners. They were Jessica Wigmo and Forced Upon Us, and the reason I liked them was because they both had winning form over the course and distance and were both drawn high which was likely to be an even greater advantage in this big field.

In fact Jessica Wigmo had the rails draw which had a great record over seven furlongs here, and although Forced Upon Us had a four-month absence to overcome, I was willing to give him the benefit of the doubt.

As the race was approaching I happened to mention to my wife that, as far as I was concerned, only two horses could possibly win to which she responded, "but they can't both win!"

I started to think how right she was and that staring me in the face was a way in which if my judgement was right, I could capitalise on this opportunity on all fronts. I had not used the exacta bet all night and with only one race left, this was my last chance. I backed Jessica Wigmo at 10-1 and Force Upon Us at 7-1, whilst pairing them in a reverse exacta.

The race was soon under way and although there were still several horses in a heap entering the last furlong, my two selections were right in there amongst them and it was with great delight that I saw Jessica Wigmo go past the line in front, just in front of Forced Upon Us. Backing a nice-priced winner was good, but the exacta was even better as it paid £91.60 to a £1 stake.

I have never been a great fan of the term "getting out stakes" for the last race at a meeting as I don't believe that your staking should be determined by the position of a race on the card. I consider all of the meetings that I attend to be one big ongoing fixture and I like to think that my bets on the last race would have been just the same had it been the first on the card, but I must admit that the drive home from Kempton late that night was rather more pleasurable than it might otherwise have been because of how the card had ended.

Sunday 20 April – Great Leighs

I had been in touch with Great Leighs via email for some time about the possibility of attending one of their early meetings, for which numbers were to be severely limited before the general public were allowed in towards the end of May. I really wanted to be there for the first meeting if it were at all possible, but by the time I returned from racing at Newbury the day before Great Leighs was finally due to stage its first fixture, I still didn't know whether my application had been successful or not and, by that stage, I had more or less resigned myself to the fact that I wasn't going to be there for this great occasion.

I wanted to be there for three main reasons. Firstly, my love of All-Weather racing meant that I wanted to see at first hand how the track looked and how horses performed on it. Secondly, it had been 81 years since Britain's last brand new racecourse had opened at Taunton and I just wanted to revel in the historical significance of it all, and thirdly it would mean that I had visited all 60 existing British racecourses and there wouldn't have been that many people who could make such a claim at this stage. I'm not an egoistical person, but I did like the sound of that.

When I did eventually switch on my PC that Saturday evening and browsed my emails, I had a very pleasant surprise. I had indeed received an invitation to the opening meeting the next day, so it was with great excitement that I rearranged all the plans I had made and prepared myself for one of the most significant days of my working life.

When I finally arrived at the track after a near two-hour drive that Sunday lunchtime, I had a wonderful feeling of being part of something really special. When I eventually walked across the Polytrack surface and on to the infield (which is where the stands were) it really hit me what this was all about. After a series of frustrating delays since the track was originally due to open in October 2006, this was the real thing and I was here to see it.

Once inside the stands I began to bump into some familiar faces, mostly fellow racing journalists, and we all agreed that this was a significant and exciting occasion. I then spotted a press "huddle" just a few yards away from where I was standing and I attached myself to the edge of it to see and hear what was going on. At the centre of the huddle was the course's owner and chairman John Holmes, who was fielding questions from the assorted press whilst having his photograph taken on a regular basis.

I listened very carefully to what he had to say and the one thing that struck me was just how passionate he was that this project should work. The quality of the racing was very important, but at the heart of it all was the welfare of the visitor. This wasn't just a case of building a new track, sit back and watch the pennies roll in from the "punters". Here was a man who still had a very clear vision of what he wanted to achieve despite all the adversity he had faced in the previous months and years. Having heard what he had in mind, you just wanted it to work too.

A Breeders' Cup Trial meeting was one of the ambitions, possibly tied in with a meeting on the east coast of the United States, and that sort of forward thinking just deserves to succeed. Yes, a great deal of work still had to be done on the course's infrastructure. Parts of the infield were still literally a building site (they were laying large plastic mats down with tape on the ground floor of the stands even after racing had got under way) but the paddock, the weighing room and, most importantly, the track itself were complete. The Polytrack surface looked in magnificent shape and when I wandered outside to the paddock and looked up at the strange stirrup-shaped floodlights, I began to imagine this whole area packed with thousands of racegoers. I liked what I saw.

I hadn't intended to have a bet at the meeting, as from a punting point of view this was a reconnaissance mission. I did reserve the right to have a punt if a pronounced track or draw bias had emerged in the early races, but once a few contests had taken place there were no obvious biases, which is exactly how the track's hierarchy wanted it. Horses had managed to win from the front and after being held up and it didn't seem to matter whereabouts on the track they made their efforts.

Temple Of Thebes wins the first race to be run at Great Leighs on 20 April 2008

When the fifth race arrived though, I was getting itchy fingers and decided to have a bet.

4-20 GREAT LEIGHS "ONLY IN ESSEX" HANDICAP (SPONSORED BY STAN JAMES) (0-105) (2) 4YO+			
10 furlongs		**Par = 109**	**5 runners**
AW FORM	**NAME**	**BEST**	**LAST SIX**
5/1-2	LUBERON	111L10	95K08 106K11 111L10
2/	RAYHANI	107L12	107L12
1/	CLIPPERDOWN	107L10	107L10
1/3-	ESCAPE ROUTE	103L07	94L07 103L07
44/	SNOQUALMIE BOY	95L08	95L10 95L08

The reason I was attracted to this race was that according to my own race card Luberon was the only guaranteed front-runner, and he had the best previous ratings on sand as well. He was also fit from a decent return to action at Lingfield earlier in the month, but despite all of that he was on the drift in the market out to 4-1. With one-quarter the odds a place, I thought he was a cracking each-way bet at that price, as I couldn't imagine that two horses would be able to get past him if he was able to get to the front, which I was sure he would.

Sure enough, he did enjoy an uncontested lead and I was sitting pretty for most of the race, but unfortunately for me Escape Route, whom I had stamina doubts about, came and did him in the latter stages. He never looked like finishing worse than second though, so nothing was lost.

I waited until the last race before going in again.

5-20 FIRST SINCE 1927 HANDICAP (SPONSORED BY STAN JAMES) (0-60) (6) 4YO+

10 furlongs **Par = 101** **13 runners**

AW FORM	NAME	BEST	LAST SIX					
-8564946	BIENHEUREUX	107W12	95L12	91K11	99L13	84L13	102L16	88S12
8-532512	SIENA STAR	107L10	103L10	93K11	105L10	91K11	106L10	98K10
73-52780	SORBIESHARRY	106S08	94S08	90S11	94K10	99K10	91W09	82S11
0300-322	KINGS TOPIC	106L10	100L10	80W08	52K10	102L10	106L10	91L10
54493-25	MAGIC AMIGO	104K12	93K10	100K10	82K11	100K12	89K12	96K10
43177626	TABULATE	103L08	102K10	92K08	85W09	94W09	96W09	96K10
751-	SIR DUKE	102W12				94W09	97K12	102W12
0090/19/	SENOR SET	101W13	82W12	75W13	83S11	75S16	101W13	97L16
-3957344	ROYAL EMBRACE	100K08	98W07	89K08	94K08	92W08	87W08	89W08
-8954111	BAHHMIRAGE	99W09	66W09	90W08	97K08	90W09	92K10	99W09
060/0/22	RULING REEF	99S12	77L07	88L08	85W09	80K10	98S11	99S12
57/8-	BLOCKLEY DJALALABAD	87W08				81W07	87W08	85W09

Siena Star was at his best at this trip on a left-handed track and 10-1 looked generous under the circumstances, whilst Kings Topic looked another each-way bet to nothing at 4-1 as conditions seemed likely to suit him too.

Siena Star ran well for a long way, but never looked that happy even though he was going the right way round. He eventually dropped out, but despite being forced very wide throughout Kings Topic managed to plug on down the outside to finish second and save the each-way money, albeit a long way behind the winner Sir Duke.

I went home buzzing and I couldn't wait to go back. This had been a great experience and one that I shall keep with me always. The rules regarding opening a brand new racecourse mean that Great Leighs cannot stage races above Class 2 level within its first year, but this track deserves to have top-class horses running on it and when the project all finally comes together, I'm sure that will happen.

Monday 12 May – Wolverhampton

This meeting marked the first day of a week's mini-break away from the trials and tribulations of everyday life. Horseracing may be my profession, but despite being exposed to it day-in day-out in my job for over 11 years now, happily it's still my hobby also.

This was also the only All-Weather meeting that I would be attending during the week, so my betting would be on a totally different scale today.

No-one could possibly pretend that this was anything but a very mediocre meeting, even for one on sand. Indeed the following day's *Racing Post* would carry a very scary headline:

"PUNTERS HIT BY HIGHER MARGINS ON ALL-WEATHER – DONOUGHUE"

with a sub-headline of:

"Small fields blamed for bigger overrounds".

Lord Donoughue was chairman of the Starting Price Regulatory Commission, which had concluded that smaller fields had helped cause a rise in the overround per runner (the theoretical profit in favour of the bookmakers built into each horse's price). This had apparently caused bookmakers' profit margins on SPs to rise by an average of 14.5 per cent.

Increasingly small fields on sand were becoming ever more noticeable to everyone, including me, and it had been suggested that this was down to too many meetings taking place. Certainly things were very different to when I first started getting really interested in this part of the sport around 15 years ago. In those days, there would be just one All-Weather meeting each day between Monday and Saturday throughout most of the winter, but now there are regularly two and sometimes even three fixtures taking place on one day. Even to a simple soul like me it seems obvious that so much racing must stretch the All-Weather horse population to breaking point.

Staring across the empty stretch of concrete in front of the grandstand at the four bookmakers on duty, I had a strange mix of emotions. It was a lovely day, very tranquil and very peaceful, in fact the sort of day you are just glad to be alive. However, the fact that it was so peaceful because hardly anyone was there could be seen as a double-edged sword, and I began to fear for the future of All-Weather racing if this situation continues. Part of my job is to encourage anyone that has not yet been to an All-Weather meeting to at least try it. If you then find that you don't like it then fair enough, but you might just find that you do.

From a selfish point of view, such an increase in the volume of All-Weather racing makes it increasingly harder to keep on top of things and a greater and greater amount of my own time is taken up in doing so, but to me it's still a labour of love and all the research and hard work does have its rewards, such as unearthing a nice gem at this particular meeting.

The first three races on the card were all over five furlongs. The opener was a maiden fillies' event and the second was a claimer, but both races were confined to two-year-olds and with so little previous form to go on especially on sand, I decided to just watch them with no financial interest. Both races did, however, provide me with an enormous clue as to the likely outcome of the third race, which was another claimer for older horses.

3-10 HORIZONS RESTAURANT CLAIMING STAKES (6) 3YO+								
5.1 furlongs		**Par = 101**				**6 runners**		
AW FORM	**NAME**	**BEST**	**LAST SIX**					
21308-31	DESPERATE DAN	107S05	104L05	101L05	94L06	90S06	95W05	100W05
13611213	DODAA	107L05	89W05	103W05	105L05	96W05	107L05	96S05
51543-01	HAMAASY	105S06	102S06	89S05	99S05	105S06	87W06	99S06
89-42385	NIGHT PROSPECTOR	105S06	95W06	94L05	100L06	96L06	92L06	89W05
-3324425	LADY HOPEFUL	103W05	91K05	90K05	90W05	92K05	97W05	86W05
8864211-	LADY BAHIA	103K05	95W05	97W05	90S05	99L05	101S05	103K05

The clue was that in both the first two races the winner had made all, so my immediate aim was to try and find the most likely front-runner in this race. Thanks to the data at my disposal in my own All-Weather race card, there was absolutely no doubt who was going to set the pace and the bonus was that it was one of the two top-rated horses on my figures.

It becomes clear why I was getting excited about this race when you look at the All-Weather form for our old friend Dodaa so far in 2008.

09JAN2008	WOLV	5.1	VS	42	7CLF	1	12	9-0	3	L -9	97
18JAN2008	KEMP	5	SD	57	7CLF	1	6	8-9	8	L +5	97
26JAN2008	WOLV	5.1	SW	58	6HCP	3	9	8-9	5	L +5	96
08FEB2008	WOLV	5.1	SW	42	6HCP	6	6	8-1	1	L -19	89
21FEB2008	WOLV	5.1	SW	64	6HCP	1	9	8-7	5	L ---	103
03MAR2008	LING	5	SD	53	6HCP	1	8	8-6	3	L -4	105
22MAR2008	WOLV	5.1	SD	56	6HCP	2	7	8-9	3	L -5	96
27MAR2008	LING	5	SD	44	6HCP	1	9	8-4	5	L -8	107
04APR2008	SOUT	5	FT	75	6HCP	3	13	9-1	5	L ---	96

The column to concentrate on is the third one from the right, which shows that the gelding had led early in every single race on sand so far this year and he had done so under a variety of circumstances, i.e. on both Polytrack and Fibresand, on left-handed and right-handed tracks and in fast run and moderately run sprints, so he was the archetypal trailblazer in every sense of the word. He was always likely to lead whatever was up against him, but the withdrawal of a horse called Gross Prophet was still a bonus as it removed the only horse that may have taken him on. Things were looking better and better all the time.

Despite having so much in his favour, he wasn't even favourite for this. That honour fell to Desperate Dan who had been a real character during his career having failed to deliver as much as he had promised in many of his races. In fact I had been doing the analysis and report for the *Racing Post* when the gelding was picked up for 12,000gns by trainer Andy Haynes after winning a seller over this course and distance in August. I had questioned then whether his new yard could get much more out of him, but to be fair he had won a couple of races since then and he would have been 8lb worse off with Dodaa had this been a handicap.

Watching Desperate Dan leaving the paddock suggested he still had a few quirks, as he seemed a bit reluctant to make his way out on to the track. His connections even tried placing a blanket over his head in order to try and disorient him so he would then consent to move forward. This wasn't exactly a case of a horse misbehaving seriously, but it was still a bit of a negative and in my mind it made Dodaa an even stronger fancy. The bookmakers weren't taking any chances as Dodaa was on offer at 2-1 and he steadfastly refused to budge from that price, but he still looked a bit of a shoo-in to me, so I had to take it.

Everything went perfectly for Dodaa right from the start. He pinged the gates and although he didn't have much to spare over his rivals by the time he hit the line, he never really looked like getting caught. To be fair to Desperate Dan, he would have finished much closer had he not met trouble as he tried to squeeze through against the rail inside the last furlong, but I did call it correctly as far as the shape of the race was concerned and all the clues were there if the relevant data was at your fingertips.

The next race on the card was a handicap over just short of one mile six furlongs.

3-40 WOLVERHAMPTON-RACECOURSE.CO.UK HANDICAP (0-65) (6) 4YO+

13.9 furlongs **Par = 101** **12 runners**

AW FORM	NAME	BEST	LAST SIX					
119166-6	ADAGE	107W13	107W13	96K16	101W13	101W13	97L16	98L13
153-1	ME FEIN	105S11			98L12	81K11	97W12	105S11
39-88770	THREE THIEVES	104K16	54W12	89L10	83L12	99W12	83W12	94K12
0/769-93	BOBSLEIGH	103K16	94L16	98K16	96K16	96L13	97L13	100K16
03/7630-	SWORDS	102W12	96W16	102W12	99S14	97W12	100W13	82K12
8/6-0341	SOVIETTA	102K12	93K12	101K16	70S11	98S16	93W16	102K12
45356-08	CAVENDISH	100W12	91W09	97K12	96W09	96L12	80K11	94W12
5636-310	TIOGA GOLD	100S14		98S12	96S14	94S14	95S14	72S14
60/104-6	HUGS DESTINY	99W13		96W12	98L12	94W13	99W13	76W13
183960-0	FINISHED ARTICLE	99W08	86W12	85W12	96L16	98W13	88K16	51K16
6	SALOON	93W16						93W16
6-	LADY PICKPOCKET	90S11						90S11

I liked the look of Adage as he had some good form over this course and distance. In fact he had raced over it four times in his life and won three of them, so he just had to be backed at 6-1.

My eye was also attracted to Me Fein because of that last run and when a horse owned by Barney Curley is backed in from 3-1 to 13-8 the alarm bells should have been ringing very loudly in my ears, but I had discarded him because he seemed to improve quite a bit for the switch to Fibresand and in my mind there was no guarantee he would reproduce that last run back on Polytrack.

I was wrong, Me Fein won easily whilst Adage finished a never-dangerous fourth. In hindsight I should have backed the winner as well before the price started to collapse. After all he had previously won on Polytrack, so I was probably a bit harsh in dismissing him so readily.

The penultimate contest was a handicap for three-year-old fillies over the extended mile.

4-10 NAME A RACE TO ENHANCE YOUR BRAND FILLIES' HANDICAP (0-70) (5) 3YO

8.6 furlongs **Par = 96** **11 runners**

AW FORM	NAME	BEST	LAST SIX					
5-4	QUEEN'S SPEECH	96K08					94W08	96K08
3-23	DREAM SEA	95K08				89K07	95K08	84L10
09242-4	SUNSHINE LADY	93W08	75W06	78W06	93W08	85W08	88W08	90S08
421-	OGRE	93K08				87W07	91W08	93K08
6-3	STAR GRAZER	90W09					84K07	90W09
32-	PHARAOHS QUEEN	90L07					90L07	88K08
63-226	INONTIME	88S07	85W06	85W07	88S07	87S07	86W08	
0- I	TALIAN GODDESS	85L07						85L07
6-	RECLAMATION	82L08						82L08
07-	FLOWER	82K08					82K08	81L08
8-	DOUBLE ON RED	82K07						82K07

I was attracted to Queen's Speech, but was a little concerned that a wide draw in stall 11 wasn't exactly ideal. With such a short run to the first bend it's better to be drawn lower, but 8-1 was a big price for a John Gosden-trained filly in a race like this so I waded in before going down to the start which is almost in front of the stands.

To be fair to Robert Havlin he did everything right aboard Queen's Speech once the race was underway, especially as the combination were forced to race wide throughout because of the draw. The filly even led for a few strides starting up the home straight and was still just about in front passing me, but Sir Mark Prescott's Reclamation proved a very determined rival and rallied to worry Queen's Speech out of it by a neck.

The last race on the card was for amateur riders which isn't usually the sort of race I like to get involved in, but two in the field stood out above the others as far as I could see and they were on offer at prices that would produce a profit if either of them won.

HOTEL & CONFERENCING AT WOLVERHAMPTON RACECOURSE AMATEUR RIDERS' HANDICAP (0-55) (6) 4YO+

12.2 furlongs **Par = 101** **12 runners**

AW FORM	NAME	BEST	LAST SIX					
61541365	LORDSWOOD	108L10	93K08	96K08	108L10	107L10	84L13	87L13
8-41734	STARSTRUCK PETER	105L16	94D10	96W13	100W16	105L16	102K16	0G16
71878-65	THORNY MANDATE	104W12	100W12	96W13	93W12	97W12	97W12	104W12
539-3666	TREETOPS HOTEL	104L12	88W12	88W12	92W12	89W13	85W09	86W12
17242-00	PRINCE OF GOLD	101W06	90W07	94W07	100W07	95W07	91W07	84W07
257-1316	CRAGGANMORE CREEK	101S12	94W12	98W13	83S16	93S12	92S11	96S14
35/6-	FAIRLY HONEST	96L10				90L07	93W08	96L10
5/8-	MONSIEUR DUMAS	95W09					88W08	95W09
225/0	WEE ELLIE COBURN	95W06			89S06	95W06	86S07	77S08
6/7/7/10	WIZARD LOOKING	94K12		91L08	84L08	91L08	94L12	94K12
540080-	GATECRASHER	91K16	85W12	83K12	91K16	88W13	82W13	64W13
80/	REVOLVING WORLD	84L12					84L12	55S12

Starstruck Peter and Thorny Mandate had both been successful at this track within the past year and both had earned decent ratings from me recently (I hadn't been able to assess Starstruck Peter's last run at Great Leighs due to a lack of data). I backed them at 3-1 and 4-1 respectively and was sitting very pretty when both of my selections were in a group of three that had broken a long way clear of the others on the home turn. Unfortunately, the only one of the trio I didn't back, Wizard Looking, proved too good for both of them.

Things had looked so promising after Dodaa had won and having talked myself into believing he had such an outstanding chance, I really should have had more on. I couldn't build on his victory nonetheless, so I just had to concentrate on enjoying the rest of my holiday.

Wednesday 28 May – Great Leighs

Great Leighs was now ready to open its doors to the general public and I was very interested to see how much things had changed since that opening meeting more than five weeks earlier.

The fixture very nearly didn't take place at all, however. It was announced early in the morning that there would be an inspection before racing could be given the go ahead, as work requested by the BHA hadn't yet been carried out with regard to the judge's box and access for emergency vehicles. It had been very wet the previous weekend which had apparently thwarted attempts to get the work done in time, but thankfully those involved worked very hard to remedy the situation and the green light was given well before I was scheduled to start the near two-hour journey up to Essex.

Having negotiated the animal rights protesters outside the main gate and parked the car, the first thing that struck me was the number of other vehicles and indeed the number of people walking past me. This place now had the look and feel of a proper racecourse and I was looking forward to seeing what else had changed since April 20th.

There was still plenty of mud and even after walking across from the car park my shoes were very quickly covered in the stuff. They had certainly had some rain here, but as far as the infrastructure was concerned the improvements were truly remarkable and it's worth remembering that they had staged five more fixtures with just an invited audience since the inaugural meeting.

The inside of the grandstand had been transformed and was a much more inviting place now, whilst outside there was a separate area for the bookmakers which had much more the look and atmosphere of a traditional betting ring. There were plenty of stalls serving food and drink behind the stands and the area between the stands and the paddock had much more hard ground to walk on rather than the swamp which had been there before.

There was certainly still a huge amount of work to be done, but they were getting there and I was still convinced that when it was all complete this would be a fantastic facility.

As far as the actual racing was concerned the six-race card was an attractive one, but I was to make too many mistakes on the day and if there is any value in going through them explicitly, it will at least demonstrate the pain resulting from poor judgement. In other words, don't try this at home.

The first mistake was in getting involved at all, as I still didn't have enough data from Great Leighs to be able produce any meaningful data over the draw or running styles and I still wasn't yet able to calculate my own speed ratings for races that had taken place here. Despite all of that, I still went ahead and had a punt in the first race, but quickly paid the penalty.

2-30 LOOKERS LAND ROVER HANDICAP (0-85) (4) 4YO+

5 furlongs **Par = 105** **8 runners**

AW FORM	NAME	BEST	LAST SIX					
/2854-08	RUSSIAN SYMPHONY	109L06	102L07	97K07	101W06	103L06	100L07	97K06
46625-13	HAMMER OF THE GODS	109L05	94S05	100L07	102W06	99K06	104W06	91W06
/1457-69	DIANE'S CHOICE	108K06	103K06	103L06	106L06	102L06	99K05	99L05
38133335	ALMATY EXPRESS	106W05	104W05	100W05	104W05	104W05	0G05	93S05
11-23676	SILVER PRELUDE	106L05	106W05	99L06	106L05	96L05	101W06	98L05
6655-055	NEW YORK OSCAR	104W05	103L06	99S05	87S06	79S05	97W05	0G05
-5136311	LORD OF THE REINS	103K05	103W05	100S05	90L06	99K05	103K05	0G05
	ICE PLANET							

Several people that I had spoken to or that I had heard discussing the contest beforehand, had come to the same conclusion as me, i.e. that there would be plenty of pace on in this. Several of these horses had made the running on at least one occasion in their careers and both Almaty Express and Silver Prelude were established trailblazers. Some people had therefore concluded that the front-runners would cut each other's throats and leave it open for a closer, but I wasn't quite so sure.

When I had come across this situation in the past, things hadn't always quite worked out as I thought they would. Quite often one or two of the confirmed front-runners would change their running styles and be ridden more patiently. Perhaps their connections had looked at the race and decided that it wouldn't be in their best interests to take on the other pacesetter(s). The result often was that one horse would end up with an uncontested lead and go on to make full use of it.

So the question was, which one might enjoy that scenario here? An even bigger question was, why the hell did I force myself to back only one horse? The odds available on both of the front-runners made them tremendous value if one of them did get a soft lead, but for some reason I came down on the side of Silver Prelude who was available at 10-1. I think the reason was that Almaty Express had already raced once at Great Leighs before and he hadn't managed to lead despite trying to, eventually finishing third behind Lord Of The Reins. I just couldn't get that out of my mind.

Anyway the inevitable happened. Almaty Express pinged the gates, soon established a significant lead and never looked like getting caught. My pig-headedness had cost me a 9-1 winner.

The second race was a two-year-old maiden with very little form to go on, so I was happy to sit the race out and just watch, whilst the third race was a nice conditions event where only two of the five runners had raced on sand before, one of them just the once almost three years ago. That made it another race to watch rather than get involved in, though it did produced a cracking finish with Mutajarred getting the better of the favourite Many Volumes by a neck following a protracted battle.

One thing that I had noticed about the three races that had taken place so far was that all the winners had come two or three horse-widths wide into the home straight, and indeed at most of the previous meetings here it did seem to be an advantage to make your run down the middle of the track rather than stick to the inside rail. So having made a note of that, why on earth did I back a horse drawn in stall one in the fourth race?

4-10 ESSEX RACING CLUB HANDICAP (0-90) (3) 4YO+

6 furlongs **Par = 107** **9 runners**

AW FORM	NAME	BEST	LAST SIX					
390/33-4	ROYAL ISLAND	112L10	112L10	100W08	103K10	106D07	104D08	105D08
3914-122	SWEET PICKLE	110S07	97W06	103W06	97W06	104S06	109S07	110S07
34133315	CEREBUS	108S07	101S05	104S06	100S06	0G06	0G06	97S06
68-21124	FROMSONG	108K06	98L06	106L05	107K05	108L06	107L05	108K06
92-31111	CAME BACK	107S06	100S05	99W06	99S06	107S06	101L06	106S06
5/	OBE GOLD	105K06						105K06
362-	RAINBOW MIRAGE	100W07				98W06	86L06	100W07
/0617-25	TILLY'S DREAM	100W06		78S06	96W06	91W06	99S05	98S05
37-7	MANGO MUSIC	96W06				96W06	94L06	90L07

The three I had focussed on were Sweet Pickle, Came Back and Fromsong. This was another race where I should really have backed more than one horse, but I was still nervous about going too heavily at a new racetrack so again I forced myself to back just one.

I discarded Sweet Pickle as I considered her a better horse over seven furlongs on Fibresand, even though she had won over this trip on Polytrack. I also ruled out Came Back because I thought he was also better on Fibresand and as yet I had no evidence that form shown on that surface could be successfully transferred to Great Leighs, but that was my biggest mistake. If I had looked deeper (and I did have his entire record on sand with me) I would have seen that his form on Polytrack was every bit as good as it was on Fibresand.

Instead I backed Fromsong at 12-1 on the basis that he had been running really well at a couple of the other Polytrack venues lately, and I hoped that he would be tucked in from his rails draw and then be switched wide to make his effort from the home bend. Unfortunately, he never got off the rail and could only manage eighth of the nine runners behind Came Back. I had now missed out on a 7-1 winner.

The fifth race on the card was quite a valuable fillies' handicap and I had narrowed it down to two, but that became one when one of the pair was announced as a non-runner.

4-45 IKANDI INTERIORS FILLIES' HANDICAP (0-95) (3) 3YO+

10 furlongs **Par = 104** **8 runners**

AW FORM	NAME	BEST		LAST SIX				
/11-3565	LISATHEDADDY	109L10	108L10	109L10	97L10	108L10	102L10	101K11
28-113	CAPE VELVET	105W08		105W08	86W08	87W08	98W08	91W08
3870-411	NAUGHTY THOUGHTS	103L08	93W09	87L07	89L10	91W09	97L08	101W09
1/4	FOLLY LODGE	100K08					88K07	100K08
1	DANCING ABBIE	99W09						99W09
1-0	FONGS GAZELLE	97K11					97K11	0G10
1-	DUSTY MOON	96L07						96L07
1-	ZASKAR	94W06						94W06

Lisathedaddy looked a different class to these, at least on sand, and so 6-1 looked a fair price to me. There were some unexposed three-year-olds in the line-up though, each of them having won their only starts on sand so far, so I was rather more hopeful than confident. My fears proved accurate as Zaskar and Dancing Abbie filled the first two places with Lisathedaddy only fifth.

I decided to try and blast my way out of trouble in the closing three-year-old sprint handicap by backing a couple.

5-15 W20.NET HANDICAP (0-75) (5) 3YO

5 furlongs | **Par = 99** | **9 runners**

AW FORM	NAME	BEST	LAST SIX					
-1231316	ORPEN'S ART	103S05	96K05	95S05	90L05	95K06	97L05	88W05
82203441	MAGGIE KATE	101L06	101L06	88K07	94L06	89L06	0G06	0G05
54221-15	YANKEE STORM	98S06	93W05	90W06	88W06	90S06	98S06	90S06
515-4	WAVERTREE PRINCESS	97L05			92L07	97L05	95L05	90K05
3-1	THE MAGIC BLANKET	97K05					91K06	97K05
5148-	FITOLINI	93W06			66S05	93W05	93W06	87W06
30-8	L'ART DU SILENCE	91L06				91L06	60K07	70S06
74524-	REPLICATOR	89L06		65L05	79K05	88L06	89K06	89L06
	QUAROMA							

I decided to stick to the top two on my ratings, both of whom should have had no problems with the conditions. The only other one I had considered was Yankee Storm, but I considered him to be a Fibresand horse despite the folly of such thinking earlier on the card. He was likely to find this trip too sharp anyway and I was more than happy to see him completely miss the break and give his rivals a start. There was surely no way back for him now, was there?

I backed Orpen's Art at 16-1 and Maggie Kate at 9-2 and was therefore getting quite excited when Orpen's Art was brought through to lead under the very inexperienced 7lb claimer Holly Hall a furlong out, but wait! What was this finishing fast down the outside? Of course it was Yankee Storm who had engaged the afterburner once switched out wide and swept past late to snatch my (substantial) winnings from me. No blame at all could be directed at Ms Hall, who rode a fine race.

Any blame could only be directed at me, as I had been rather quick in dismissing horses that had shown good form on Fibresand and I had paid dearly for it. There is a vast difference between horses that have shown *slightly* better form on Fibresand than on Polytrack and horses that had shown *substantially* better form on Fibresand than on Polytrack. Both Came Back and Yankee Storm very much belonged in the former category.

This was a day when my betting bank had taken a battering and it damned well shouldn't have done!

Friday 4 July – Southwell

I wished that I had been able to get to Southwell before now for the purposes of this diary – my last visit had been in January – but a combination of circumstances and a busy schedule meant that this was my first opportunity. It was great to be back though, and what was even better was that I liked the look of the six-race card too.

The first race was one of my favourite types of race run on sand, a handicap over the Southwell straight five furlongs.

2-30 CARTWRIGHT KING HANDICAP (0-65) (6) 3YO

5 furlongs **Par = 98** **13 runners**

AW FORM	NAME	BEST	LAST SIX					
-4223244	JALONS BRIDEWELL	101L06	94L06	93S06	101L06	92W05	83S06	96W05
31437	NOTE PERFECT	98S06		86S06	91S06	88K06	98S06	84S07
-6151453	MAC DALIA	98S05	90S05	83S07	95S05	74S06	87S05	91G05
4-384634	MUJAHOPE	97L10	93W09	84L10	97L10	81K08	72W08	88S07
4432-05	MYSTICKHILL	96W05	82S05	90W06	93W06	92W06	71W06	96W05
7084-50	WESTWOOD DAWN	89W05	62W06	71K07	71W07	89W05	85S05	85G06
7-3	CARMINE ROCK	88S06					84W05	88S06
04-777	SANTA CLARA	86S06		86L07	85S06	84S07	86S06	72W06
005	WICKSY CREEK	82L06				54L08	82L06	80G05
00-	MR FUNSHINE	80W05					69K06	80W05
8-9	LADY AVIATOR	70S05					70S05	50S07
0-	ABITOFAFATH	64W06						64W06
	ACCLIMATE							

There was only one thing I was sure of in this particular race. I wanted a horse with early pace and drawn low, preferably between one and five, as they were the stalls that had been doing especially well in recent months in terms of percentage of winners to runners.

My eye was immediately drawn to Mac Dalia, who was three from five over this course and distance. Admittedly her wins were all sellers, but this wasn't much better and she would have no problem taking an early prominent position looking at her style of running. What's more she was perfectly berthed in stall three, so she just had to be backed.

There was a time when I would have had a few quid on the horse drawn in stall one over the straight five furlongs here, provided the animal in question had shown a little early speed at some point in its career, but although the statistics for stall one were still good they weren't as dominant as they had once been. Moreover, the horse berthed there, Mr Funshine, hadn't shown anything like enough ability to justify any kind of support, just as his long odds would have suggested.

The horses drawn in stalls two and four had nothing to recommend them either, and although the horse in stall five, Carmine Rock, had shown a little ability on the surface in her last start, that had been when held up over six furlongs, so under these conditions she looked poor value at around 4-1. I was now getting more and more confident about Mac Dalia, so I had something of a strut in my step when I waded in to take 5-1 about her.

Everything looked rosy for a long time once the race was under way too, as Mac Dalia pinged the gates and bowled along down the favoured centre of the track in front, but what was this sat on her shoulder going just as well? It was no less than Mr Funshine, who then proceeded to breeze past my selection and win going away at odds of 33-1. What a sickener!

I have never been that comfortable backing horses blind even though I have occasionally done it when my own statistics have recommended such a course of action. Had this race taken place even a year ago, I know I would have backed the winner.

I was feeling a bit battered and bruised and normally I wouldn't have touched the following race, a mile and a half claimer, with a ten-foot barge pole. However, this was an exception and for a good reason.

3-00 LINCOLNSHIRE DEVELOPMENTS LTD CLAIMING STAKES (6) 3YO+

12 furlongs **Par = 101** **12 runners**

AW FORM	**NAME**	**BEST**	**LAST SIX**					
56-27218	NIMELLO	114W08	97S12	97S12	81W12	98S12	100S12	84S12
3/22-422	BLACK FALCON	109S11	102S08	104S12	109S11	107S11	107S11	94S12
5035	SHANDELIGHT	107W12			92S12	91K10	107W12	97S11
23-81322	TURNER'S TOUCH	106L10	91K11	100K12	97L12	82K12	95K12	101K11
13432374	STARCROSS MAID	103S12	98S11	90S11	90S11	88S11	90S12	88S12
0-77142	PAUL THE CARPET	100W12	79L08	85S08	92S11	100W12	94W12	96W12
02/3680-	NORTH WALK	99W08	94L08	99W08	87L12	96W12	92W08	92W08
/2657237 M	ATINEE IDOL	95S11	91S12	90S12	90S14	76S12	92S14	91S14
66-334	PERSISTENT	94W09		83L07	87S07	93W09	94W09	92S08
60	TOON ARMY	88L12					86L12	88L12
0-	SIBO BAGGINS	68K07						68K07
	BRATHAY							

There was no getting away from the claims of Black Falcon. Admittedly he had broken a losing run stretching back to November 2004 when winning on turf at Beverley the previous month, but at least he had won, so he was coming into this race in form. A glance at his speed ratings showed that on his day he was well above this class and he had a string of high ratings in his recent record, not just one as with some of his main rivals.

He was currently available at around 6-4, not a life-changing price in a race like this, but it became tempting when looking at the prospects of his main rivals. Shandelight had seemed to find this sort of trip too far on this deep surface in the past and was only as short as 4-1 because she was coming into this off the back of an easy success on turf.

Turner's Touch was an even bigger enigma. He had the ability, but he was quirky to say the least and was really a Polytrack specialist. Also he had failed to show anything at all in two previous tries here, probably because his hold-up style meant that he got Fibresand kicked in his face and he didn't like it.

So I took the 6-4 about Black Falcon and I have to admit that at no stage during the contest did I feel I wasn't going to win. The money was safe from a long way out and I was back in front on the day.

The next two races were both maidens, one for older horses and one for two-year-olds, and with very little to go on I was happy to sit them out.

The fifth race on the menu, a one-mile handicap for three-year-olds, was a different matter.

4-45 ROSEMARY HERON MEMORIAL HANDICAP (0-65) (6) 3YO

8 furlongs **Par = 98** **12 runners**

AW FORM	**NAME**	**BEST**	**LAST SIX**					
7-37	LITTLE TOTO	96W08				92K08	96W08	85K07
6	MR FANTOZZI	92K08						92K08
8-5	DOUBLE ON RED	91W08					82K07	91W08
30-86	L'ART DU SILENCE	91G05			91L06	60K07	70S06	91G05
64306	REPRIEVED	90S11		76S08	89S11	90S11	89S11	78S07
1	SO SUBLIME	90S07						90S07
8-	SHEIK'N'KNOTSTERD	87L07						87L07
6	CHEEKY CHILLI	86S07						86S07
5-	SPECTRANA	84K08						84K08

U	MISS OLIVIA	0W08						0W08
	RIDE A WHITE SWAN							
	HIGHLAND HOMESTEAD							

Normally I would have backed Little Toto had it not been for that last run. Even allowing for the shorter trip it was too bad to be true and I couldn't find a valid excuse, so with the possibility that he was either going the wrong way or had a problem, I gave him the swerve.

The two I liked were Mr Fantozzi and Double On Red. The former had run a bit better in his only previous outing on sand than his sixth placing might suggest in a race that had already produced winners, and he had also been performing well on turf recently, especially in the context of this race. He was currently available at 8-1 and that looked good to me.

The other one I decided to back was Double On Red, who ran much better on her second sand outing than on her first – usually a good sign – and she had also been an eye-catcher in that Wolverhampton contest. She was being backed in the ring, so I took the hint and snapped up some of the 9-2.

The race went perfectly. Mr Fantozzi was in front after a furlong and for a long way it seemed as though he and Double On Red would have the race to themselves. Double On Red eventually weakened away into third, but Mr Fantozzi was never in danger of defeat and romped away to win unchallenged. This was turning into a good day.

A pattern had also emerged. All five winners at the meeting had either led early or been close to the pace, and in fact only one of the other ten placed horses had come from off the pace. Early speed was definitely holding up and it was with that in mind that I went about tackling the last race on the card, a seven-furlong handicap.

5-15 SHADES SCREEN PRINT LTD HANDICAP (0-70) (5) 3YO+

7 furlongs **Par = 103** **13 runners**

AW FORM	NAME	BEST	LAST SIX					
10411-08	DOCTOR'S CAVE	108S07	72S07	98W07	108S07	105S07	45S06	78W07
26759660	MISTER BENJI	106S08	93W07	86W08	96W08	97K08	96W07	93W07
13225-43	SHERIFF'S SILK	106S06	103S07	106S06	98S06	100S07	88S07	99K06
25184462	MY MENTOR	106L10	100S12	82L12	98S12	98G14	100W12	104L10
11310-89	MEGALO MANIAC	105S07	98S07	105S07	98W07	80W06	90S06	79S06
11726421	ELUSIVE WARRIOR	105S07	95W07	92S08	95S06	100S08	97S07	102S07
/9116-31	DANCING DEANO	103S07	95W06	96W07	96S07	86S07	96W07	103S07
-8000736	UNION JACK JACKSON	101S08	66S05	60S07	81W08	91S06	90S06	87S07
555-9889	AUTOGRAPH HUNTER	100K12	97K12	97W12	88K12	82W09	79K12	89K11
1/29-711	SULARNO	99S08	89S07	99S08	87W12	90G14	99S08	95G08
92149009	GOVERNMENT	97S07	95S07	86S07	62S08	74S07	75S06	75S07
00/9-6	SURWAKI	97K07			95K08	97K07	94S08	96G08
3321 Y	OUNG GLADIATOR	92S06			91W06	88W06	92S06	92S06

The one amongst the fancied horses that I didn't like was Sularno. He did have winning form over course and distance and his style of running was also fine given the evidence of the meeting so far, but unfortunately he had the outside number 14 draw. My draw statistics showed that double-figure stalls were zero from 46 over this trip in the previous 12 months and in truth his speed ratings left him with something to find in any case.

There were several others with the right credentials though, such as Dancing Deano, Elusive Warrior, Sheriff's Silk and Doctor's Cave. All four of them had winning form over course and distance and, what's more, they had earned decent speed figures in doing so. To complicate matters they also all had prominent running styles, so choosing between them, if indeed I was going to choose between them, would be difficult.

Two of them had a problem though. Sheriff's Silk was drawn in stall 12 whilst Doctor's Cave was in 13. Even their respective odds of 12-1 and 25-1 were not enough for me to ignore the evidence of my own draw stats. If I was happy enough to take on Sularno on account of his draw, then it would have been hypocritical of me to have backed either of these two.

The only other possible contender was My Mentor. He had made the running several times in the past, he had good speed figures to his name and he was well berthed in stall three. The problem was the trip as his recent outings had been over further, sometimes much further. He had even raced over as far as a mile and six furlongs as recently as May and with so much early pace in this field, I could just see him getting left flat-footed and with too much ground to make up on a speed-favouring track.

So I was happy to take 8-1 about Elusive Warrior and 4-1 about Dancing Deano. Because of the doubts surrounding two of the three co-favourites, Sularno and My Mentor, I also decided to combine my two selections in reverse exactas.

The exacta never really looked like being landed as Dancing Deano was struggling from a long way out, but equally it appeared as though I was going to bow out with an 8-1 winner as Elusive Warrior had bounded off in front and halfway up the home straight he looked to have most of the field cooked. However, my fears about My Mentor getting outpaced proved unfounded as his stamina eventually clicked in and he picked off Elusive Warrior before going on to win by three lengths.

Perhaps my selection had gone off too quick and set it up for the winner. After all, the 40-1 outsider Autograph Hunter had come from nowhere and run on past some very tired horses to finish a remote third, but in view of the way the track was riding the tactics on Elusive Warrior were the right ones. Restraining a confirmed front-runner like him would probably only have disappointed him.

The pace rating I allocated for races on the round course at this meeting was 73 whilst for the one race on the straight five-furlong track it was 83 (I tend to split races on the round and straight courses at Southwell when calculating the track variants and the speed figures as conditions can be significantly different on each). These high pace ratings show just how much of an advantage prominently ridden horses held, and that those held up off the pace could be forgiven for what may at first have appeared to be ordinary efforts.

Those pace figures would be there amongst each horse's all-in-a-line form for this meeting when they ran again on sand, like this:

MR FUNSHINE											
19SEP2007	KEMP	6	FT	52	6HCP	11	12	9-1	1	P ---	69
22OCT2007	WOLV	5.1	SD	36	6MDN	10	12	8-12	6	P ---	80
04JUL2008	SOUT	5	VS	83	6HCP	1	13	8-4	1	P ---	98

I knew that Mr Funshine had been flattered by his victory in the opener, as he had been ridden handily on a speed-favouring track (hence the pace rating of 83 in his most recent start). His next outing was on turf at Windsor, and although punters weren't sucked in by the fact that he came into the race off the back of a win (he went off at 20-1) it's worth noting that he still only managed 12th of the 13 runners, beaten over 20 lengths by the winner, and his comment-in-running was "slowly away, always behind".

Of the six winners at the meeting, My Mentor probably emerged with the most credit because –

04JUL2008	SOUT	7	SW	73	5HCP	1	13	9-10	3	H ---	101

he had been held up in his race and was the only winner at the meeting to do so.

I had ended the day in front. It could have been even better of course, but isn't that always the case?

Wednesday 16 July – Kempton

This was going to be the last meeting covered in this diary before the manuscript of the book had to be delivered to the publishers, so I was keen for it to end with a bang.

The first race on the card was an apprentice handicap over one mile and three furlongs.

6-20 WEATHERBYS BLOODSTOCK SERVICES APPRENTICE HANDICAP (ROUND 7) (0-80) (4) 4YO+								
11 furlongs		**Par = 105**			**8 runners**			
AW FORM	**NAME**	**BEST**	**LAST SIX**					
11121719	SAFARI SUNDOWNER	111L10	111L10	104W08	81L10	102K11	108K10	92G10
62436-44	ROSS MOOR	106W13	100L12	103L12	97L12	99W13	94L12	102G10
134/31-	AYPEEYES		104W12	92W08	92W08	101W09	102L10	104W12
6/2	SHOW WINNER					103K10	100W09	103K10
-1585322	ALEXANDER GURU	102W08	86K10	79L10	98W09	102W08	95K10	97K12
87/344-5	CONSTANT CHEERS	100W12	88L10	95L12	100W12	97K12	96L13	93K10
33539-	YAB ADEE		96W09	91L10	96W09	92W08	92K12	88W13
	TOWEROFCHARLEMAGNE							

I wasn't going to make this race any more complicated than it need be. Safari Sundowner had earned a couple of speed figures which were way ahead of anything his rivals could muster, and one of them had been achieved here so there were no worries about the track. With all of this in mind, I was surprised to see that he was available at as big as 13-2, so I just had to back him.

This was one of those races where everything went perfectly. After being held up in the early stages, Safari Sundowner swept around the outside of the field under William Carson and as soon as he hit the front I knew the money was safe.

A good start, but now for a very complicated-looking seven-furlong nursery.

6-50 SPILLERS NURSERY (5) 2YO

7 furlongs **Par = 94** **13 runners**

AW FORM	NAME	BEST	LAST SIX					
51	GRAND HONOUR	95W05					82K05	95W05
22	THE DIAL HOUSE	95K05					95K05	92G05
723	SWINGFIRE				94G06	73S05	94G06	84S06
2	DEAL CLINCHER						90K06	90K06
8	FASALEE						84K05	84K05
5	ENTRANCER						83W06	83W06
47	REDHEAD					81K06	81K06	78K07
8	CALYPSO PRINCE						77K06	77K06
	SCENIC PASS							
	AEGEAN WARNING							
	GASSAL							
	ANOTHER LUKE							
	BARCODE							

Nurseries at this time of year are especially difficult (they had only been going for two weeks). Previous form was thin on the ground even though these two-year-olds needed to have shown enough form in order to qualify for them. Previous form on sand was even more rare, but at least a couple of these had previously exceeded the race's class par on Polytrack and that gave me something to focus on.

The Dial House was one of the pair, and he had exceeded the class par at this track, which was a big plus in my book. The negative was that he was now trading at only around 3-1 which was short enough for a competitive race like this, but I couldn't leave him unbacked given the positives so I went in on him.

The other highest-rated horse was top-weight Grand Honour. There is an old adage that you should back top weights in early-season nurseries, but I wasn't interested in him because of that. I liked him because he was just the sort of horse whose recent efforts on turf had rather masked his earlier performances on sand. To all intents and purposes his form figures were 21448, but take away his three starts on turf and you were left with 21. It was in his last start on Polytrack when successful at Wolverhampton that he exceeded the class par for this race. It's also worth noting that his most recent outing on turf when only eighth was in a Listed race, so he was obviously highly regarded yet he was currently available at 12-1. There is no way he would have been that price had it not been for those three recent starts on turf. He also just had to be backed.

As in the first race, I was more than happy with how things went once the contest was under way. Given a copybook ride by Jimmy Quinn, Grand Honour travelled like a dream behind the leaders and the rider made perfect use of the 'cutaway' on the inside rail where the two courses meet at around the two-furlong pole. He had to battle hard to get up the inside of the leader Gassal, but he was always going to get there.

The Dial House did not enjoy a clear passage at all and at one stage he appeared more intent at taking a nip at one of his rivals rather than concentrating on going forward. Still, this was another decent result and I wanted to keep it going.

The next race on the card was a two-year-old fillies' maiden over seven furlongs. There were some interesting juveniles on show and this would be a race with the future in mind, but with virtually no form to go on it wasn't a betting contest for me today.

The next race however, a one-mile handicap, certainly was.

7-50 DIGIBET.COM LONDON MILE HANDICAP (LONDON MILE QUALIFIER) (0-70) (5) 3YO+

8 furlongs **Par = 103** **14 runners**

AW FORM	NAME	BEST	LAST SIX					
-8223175	MILLFIELD	106W09	99K07	105K06	101L07	101L08	100L08	91K07
165/935-	LITTLETON TELCHAR	106K07	104K07	98L08	106K07	93W07	102L08	97W07
6/3293-1	JEBEL ALI	104L10		101L10	99L12	94L12	103L10	100K10
2	JILL DAWSON	103L08						103L08
1-212380	WRIGHTY ALMIGHTY	102K08	96K08	102K08	97L08	95L08	95K08	94L08
046-134	GARDEN PARTY	101W09	67K08	101W09	96K08	95L08	97L08	100W08
74546173	NAPOLETANO	101L08	91L07	97K08	92L07	97L07	88L07	95L07
182-	GAZBOOLOU	101K08				101K08	99L08	99W07
6-91	KABALLERO	100L08				89S08	93G08	100L08
3310-3	AFFRETTANDO	98W07		91L07	95L06	92K07	97L07	98W07
4	IDESIA	97W08						97W08
64725	EAGLE NEBULA	96K08		76L10	90K07	82L06	96K08	88K08
67	PRINCE DESIRE	96K08					92G08	96K08
66	COPPERWOOD	91L06					90L08	91L06

Millfield was a definite contender, as not only was he joint top-rated on my figures, like Grand Honour earlier on the card he had exceeded the class par for this race at this track when runner-up here earlier in the year. He was vying for favouritism, but 6-1 still looked a reasonable price to me.

I also went in on Littleton Telchar because he had exceeded the class par here in the past (in his case more than once). There were plenty of negatives about him; he hadn't run on sand for a long time and his recent efforts on turf had been patchy, but he was on offer at 25-1 so he was worth a punt at those odds.

Things didn't start off too well for Millfield once the race got under way, as he seemed to get hampered and dropped to the rear of the field. That is less of a worry when George Baker is aboard, however, as he's at his best when bringing horses from well off the pace, so I wasn't unduly worried. I never needed to be either, as he eventually cruised into the contest in trademark fashion and when the gaps appeared at just the right time, he made full use of them to burst through and win going away.

Littleton Telchar ran an abysmal race, eventually finishing a tailed-off last, but again this was a winning race.

8-20 DIGIBET CASINO FILLIES' HANDICAP (0-85) (4) 3YO

8 furlongs **Par = 99** **9 runners**

AW FORM	NAME	BEST	LAST SIX					
3-425412	MONTIBOLI	104S08	93W05	91S06	92K06	90S05	104S08	94S07
3122-22	SMOKEY RYE	103L07	89L07	91K06	95L07	94L06	103L07	93K07
1-6	DUSTY MOON					100G10	96L07	100G10
2	DUBAI POWER						99K08	99K08
1	NICE MATIN						96L08	96L08
43	JOLLYHOCKEYSTICKS	96K10					95K08	96K10
	CENTENEROLA							
	TATBEEQ							
	BOIS JOLI							

The next on the card was a three-year-old fillies' handicap over a mile.

With all these top-rated horses winning so far, I got rather sucked in by Montiboli and Smokey Rye and backed them at 4-1 and 6-1 respectively, but I should have paid more attention to the form of Montiboli as she had clearly demonstrated a liking for a slow surface in her previous outings, including on Fibresand. Her one previous try here had been ordinary at best.

She did try her best, but was always struggling for finishing pace on this surface and could only manage third at a respectful distance. Smokey Rye never looked like winning, though it was subsequently reported that her saddle had slipped.

The race as won by Dubai Power and in retrospect I should have had something on her. She had equalled the class par for this race in her only previous outing on sand, which had been over this course and distance, but due to some modest recent efforts on turf she was allowed to go off at a very tasty 16-1!

The penultimate race on the card was an ordinary-looking handicap for three-year-olds over seven furlongs.

8-50 DIGIBET SPORTS BETTING HANDICAP (0-75) (5) 3YO

7 furlongs **Par = 100** **12 runners**

AW FORM	NAME	BEST	LAST SIX					
42234447	VALHILLEN	100G06	99W05	91K05	100G06	96G06	93W06	95K06
2-21	GREYSTOKE PRINCE				98W07	93L07	98W07	95K07
48-222	AUTUMN BLADES		98K06	90K07	91L07	94K07	66L07	98K06
2121-147	IKE QUEBEC	98G06	95L07	96W06	92W06	89K06	95K06	98G06
21	SPIN AGAIN					97W06	97L06	97W06
0-546252	TOO GRAND	96L06	90L07	90K08	87K07	77L07	92K06	90K07
6-1	REDARSENE					94W08	88K08	94W08
1-2	ROMANTIC VERSE					94W06	75W07	94W06
0-2	CONNOR'S CHOICE					89L07	69K05	89L07
0- S	TARFINCH						77K06	77K06
	OXBRIDGE							
	ANOSTI							

Valhillen was exposed and wasn't sure to stay, but he was the only horse in the field to have matched the class par. He was also well drawn and was available at 14-1, so he was backed. Greystoke Prince had shown his suitability to this track by winning over course and distance last time and was a contender on the speed figures, so he too was backed at 3-1, whilst Autumn Blades had achieved his best speed figure here last time and was currently a tempting 8-1, so money went on him too.

I enjoyed this race not only because it was ultimately a successful one, but around a furlong out the contest was being fought out between all three of my selections and it seemed just a case of which one would it be?

The answer was Autumn Blades, who just got up to beat Greystoke Prince by a head and he would have been a most unlucky loser, as he had to be snatched up when first making his move. Valhillen had every chance against the inside rail, but just tired a bit late on and was run out of third on the line.

The most important thing was that I had backed another winner and now wanted to end the diary on a real high. The problem was that the last race on the card looked about the most difficult of the night!

9-20 METROPOLIS HANDICAP (0-65) (6) 3YO+								
7 furlongs		**Par = 101**			**12 runners**			
AW FORM	**NAME**	**BEST**	**LAST SIX**					
61336007	HOLLOW JO	109K06	92K05	105K06	99K05	98K06	86K07	86K08
3324-426	MILLION PERCENT	107L05	98K07	100L07	95W07	93L07	90K08	93L08
1/2360-5	MORSE	106K06	104K06	105K06	105K06	106K06	95K06	96W06
76333280	SOVEREIGNTY	104L08	99W07	100W07	100W08	99W07	98W08	89L07
61524248	GUILDENSTERN	102K07	96W07	99K06	101W07	99W07	97W07	94K08
82111207	STRAIGHT FACE	101K07	101K07	99K07	92K07	98W07	84W07	94K08
22161913	JESSICA WIGMO	100K07		93L07	89K07	88L07	96K07	94K08
2	GRIZEDALE	100K07						100K07
7-2450	ANTHILL	99L07		84W07	94K07	99L07	94K08	92G08
0/21-347	GRAND ASSAULT	98L07	86L07	98L07	97W06	97K06	91K07	90L07
433025	MR REV	96K06	96K07	96K06	94K06	75W07	91K06	88W06
3-	QUAGLINO WAY	91L08						91L08

The top three on my ratings all had form at the track, but a couple of them seemed to have lost their way. Still, I figured that if there was a race in which they could bounce back then this would probably be it. They were all decent prices in my opinion, so I backed Hollow Jo at 20-1, Morse at 10-1 and Million Percent at 7-1.

As it turned out, of the trio only Million Percent ever got into a challenging position, but even he dropped right out to beat just one home. The other pair never figured, but if I had gone down the list to include the fourth top-rated horse then I would have backed the 10-1 winner Sovereignty.

I wasn't complaining though. The journal had finished on a high, which was something of a relief as the period outside the winter months is always more difficult when it comes to betting on All-Weather racing in my opinion. You get horses running on turf then on sand, then back to turf again and all the form becomes intertwined and messy. At least during the winter when there is no turf Flat racing taking place, the horses can only run on sand (except for the few that combine sand racing with jumping) and the form is that much more consistent and predictable. That's probably why I do better punting-wise at that time of year.

What I hope this diary has shown, however, is that if you arm yourself with a few extra tools over and above those generally available to everyone else, you can find an edge and exploit it. You can also limit yourself to following the form of a few hundred horses rather than a few thousand, and that makes the whole thing a lot easier, a lot more enjoyable and most importantly of all, a lot more profitable.

SUMMARY OF BETS

DATE	TRACK	HORSE	BET (£)	RESULT	PRICE	+/-	TOTAL (£)
Mar 15	Ling	Azabu Juban	25 EW	Unp		-50.00	-50.00
		Gallantry	50 Win	2nd		-50.00	-100.00
		Ceremonial Jade	50 Win	4th		-50.00	-150.00
		Silver Pivotal	50 Win	2nd		-50.00	-200.00
		Sgt Schultz	50 Win	1st	5-1	+250.00	+50.00
		John Terry	50 Win	4th		-50.00	00.00
		Atheer Dubai	50 Win	Unp		-50.00	-50.00
		Highland Harvest	50 Win	4th		-50.00	-100.00
Mar 28	Kemp	Fast Freddie	50 Win	Unp		-50.00	-150.00
		Lord Theo	50 Win	3rd		-50.00	-200.00
		Barry Island	50 Win	Unp		-50.00	-250.00
		Kamal	50 Win	3rd		-50.00	-300.00
		Featherlight	50 Win	2nd		-50.00	-350.00
		Mister Completely	50 Win	Unp		-50.00	-400.00
		{Featherlight/ Mister Completely}	10 Rev Exacta	Incorrect		-20.00	-420.00
		Straight Face	50 Win	1st	5-2	+125.00	-295.00
		Only If I Laugh	50 Win	2nd		-50.00	-345.00
		{Straight Face/ Only If I Laugh}	10 Rev Exacta	Correct		+243.00	-102.00
Apr 3	Ling	Kate The Great	50 Win	1st	3-1	+150.00	+48.00
		Dr Wintringham	50 Win	3rd		-50.00	-2.00
		Linda Green	50 Win	2nd		-50.00	-52.00
		Maggie Kate	50 Win	4th		-50.00	-102.00
		Salute	50 Win	3rd		-50.00	-152.00
		Bob's You Uncle	50 Win	3rd		-50.00	-202.00
		Capitalise	50 Win	4th		-50.00	-252.00
		Joy And Pain	50 Win	4th		-50.00	-302.00
Apr 4	Kemp	Angel Voices	50 Win	4th		-50.00	-352.00
		Amwell Brave	50 Win	3rd		-50.00	-402.00
		Postmaster	50 Win	2nd		-50.00	-452.00
		Grename	50 Win	Unp		-50.00	-502.00
		Gallic Charm	50 Win	Unp		-50.00	-552.00
		Jessica Wigmo	50 Win	1st	10-1	+500.00	-52.00
		Forced Upon Us	50 Win	2nd		-50.00	-102.00
		{Jessica Wigmo/ Forced Upon Us}	10 Rev Exacta	Correct		+896.00	+794.00
Apr 20	GLei	Luberon	25 EW	2nd		00.00	+794.00
		Kings Topic	25 EW	2nd		00.00	+794.00
		Siena Star	25 EW	Unp		-50.00	+744.00
May 12	Wolv	Dodaa	50 Win	1st	2-1	+100.00	+844.00
		Adage	50 Win	4th		-50.00	+794.00
		Queen's Speech	50 Win	2nd		-50.00	+744.00
		Thorny Mandate	50 Win	2nd		50.00	+694.00
		Starstruck Peter	50 Win	3rd		-50.00	+644.00
May 28	GLei	Silver Prelude	50 Win	Unp		-50.00	+594.00
		Fromsong	50 Win	Unp		-50.00	+544.00
		Lisathedaddy	50 Win	Unp		-50.00	+494.00
		Orpen's Art	50 Win	2nd		-50.00	+444.00
		Maggie Kate	50 Win	4th		-50.00	+394.00
July 4	Sout	Mac Dalia	50 Win	2nd		-50.00	+344.00
		Black Falcon	50 Win	1st	6-4	+75.00	+419.00
		Mr Fantozzi	50 Win	1st	8-1	+400.00	+819.00
		Double On Red	50 Win	3rd		-50.00	+769.00
		Elusive Warrior	50 Win	2nd		-50.00	+719.00
		Dancing Deano	50 Win	Unp		-50.00	+669.00
		{Elusive Warrior/ Dancing Deano}	10 Rev Exacta	Incorrect		-20.00	+649.00
July 16	Kemp	Safari Sundowner	50 Win	1st	13-2	+325.00	+974.00
		Grand Honour	50 Win	1st	12-1	+600.00	+1574.00
		The Dial House	50 Win	Unp		-50.00	+1524.00
		Millfield	50 Win	1st	6-1	+300.00	+1824.00
		Littleton Telchar	50 Win	Unp		-50.00	+1774.00
		Montiboli	50 Win	3rd		-50.00	+1724.00
		Smokey Rye	50 Win	Unp		-50.00	+1674.00
		Autumn Blades	50 Win	1st	8-1	+400.00	+2074.00
		Greystoke Prince	50 Win	2nd		-50.00	+2024.00
		Valhillen	50 Win	4th		-50.00	+1974.00
		Hollow Jo	0 Win	Unp		-50.00	+1924.00
		Million Percent	50 Win	Unp		-50.00	+1874.00
		Morse	50 Win	Unp		-50.00	+1824.00